Doping, Performance-Enhancing Drugs, and Hormones in Sport

Emerging Issues in Analytical Chemistry

Series Editor
Brian F. Thomas

Co-published by Elsevier and RTI Press, the *Emerging Issues in Analytical Chemistry* series highlights contemporary challenges in health, environmental, and forensic sciences being addressed by novel analytical chemistry approaches, methods, or instrumentation. Each volume is available as an e-book, on Elsevier's ScienceDirect, and via print. Series editor Dr. Brian F. Thomas continuously identifies volume authors and topics; areas of current interest include identification of tobacco product content prompted by regulations of the Family Tobacco Control Act, constituents and use characteristics of e-cigarettes and vaporizers, analysis of the synthetic cannabinoids and cathinones proliferating on the illicit market, medication compliance and prescription pain killer use and diversion, and environmental exposure to chemicals such as phthalates, endocrine disrupters, and flame retardants. Novel analytical methods and approaches are also highlighted, such as ultraperformance convergence chromatography, ion mobility, in silico chemoinformatics, and metallomics. By highlighting analytical innovations and new information, this series furthers our understanding of chemicals, exposures, and societal consequences.

2016   *The Analytical Chemistry of Cannabis: Quality Assessment, Assurance, and Regulation of Medicinal Marijuana and Cannabinoid Preparations*
Brian F. Thomas and Mahmoud ElSohly

2016   *Exercise, Sport, and Bioanalytical Chemistry: Principles and Practice*
Anthony C. Hackney

2016   *Analytical Chemistry for Assessing Medication Adherence*
Sangeeta Tanna and Graham Lawson

2017   *Sustainable Shale Oil and Gas: Analytical Chemistry, Geochemistry, and Biochemistry Methods*
Vikram Rao and Rob Knight

2017   *Analytical Assessment of e-Cigarettes: From Contents to Chemical and Particle Exposure Profiles*
Konstantinos E. Farsalinos, I. Gene Gillman, Stephen S. Hecht, Riccardo Polosa, and Jonathan Thornburg

2018   *Doping, Performance-Enhancing Drugs, and Hormones in Sport: Mechanisms of Action and Methods of Detection*
Anthony C. Hackney

2018   *Inhaled Pharmaceutical Product Development Perspectives: Challenges and Opportunities*
Anthony J. Hickey

Doping, Performance-Enhancing Drugs, and Hormones in Sport

Mechanisms of Action and Methods of Detection

Anthony C. Hackney
University of North Carolina, Chapel Hill, NC, United States

Elsevier
Radarweg 29, PO Box 211, 1000 AE Amsterdam, Netherlands
The Boulevard, Langford Lane, Kidlington, Oxford OX5 1GB, United Kingdom
50 Hampshire Street, 5th Floor, Cambridge, MA 02139, United States

Published in cooperation with RTI Press at RTI International, an independent, nonprofit research institute that
provides research, development, and technical services to government and commercial clients worldwide (www.
rti.org). RTI Press is RTI's open-access, peer-reviewed publishing channel. RTI International is a trade name
of Research Triangle Institute.

British Library Cataloguing-in-Publication Data
A catalogue record for this book is available from the British Library

Library of Congress Cataloging-in-Publication Data
A catalog record for this book is available from the Library of Congress

ISBN: 978-0-12-813442-9

For Information on all Elsevier publications
visit our website at https://www.elsevier.com/books-and-journals

Working together
to grow libraries in
developing countries

www.elsevier.com • www.bookaid.org

Publisher: Joseph Hayton
Acquisition Editor: Kathryn Morrissey
Editorial Project Manager: Amy Clark
Production Project Manager: Vijayaraj Purushothaman
Cover Designer: Dayle G. Johnson and Matthew Limbert

Typeset by MPS Limited, Chennai, India

DEDICATION

This work is dedicated to the athletes who inspired me in my youth and continue to do so now—Emil Zátopek, Elizabeth "Betty" Cuthbert, and Sir Peter G. Snell.

CONTENTS

List of Contributors.. xi
Foreword xiii
Preface... xv
Acknowledgments... xix

Chapter 1 Overview: Doping in Sport 1
History and Current State of the Problem.............................1
Prevalence by Sport, Country, Level of Competition 2
World Anti-Doping Agency and Legal Problems.................3
Ethical Issues of Usage, Sportsmanship, Character.............6
Are There Situations in Which Athletes Can
Use PEDs?—Therapeutic Use Exemption 8
Conclusion...9
References...11

Chapter 2 Anabolic Androgenic Steroids 13
Treatment of Endocrine Disruptions 13
The Athletic Drive to Increase Muscle Mass.....................14
Intracellular Biochemical and Nuclear Actions 16
Early Paradoxes and Later Confirmations in Research Findings 18
Effects of Abuse, Reversible, and Nonreversible 19
Other Anabolic Substances...21
Conclusion...21
References...22

Chapter 3 Stimulants 25
Losing Weight, Staying Awake, Gaining Focus 25
Staying Athletically Lean and on Task................................26
Neurochemistry 28
Performance Enhancement Evidence....................................30
Speeding and Crashing 32
Conclusion...32
References...34

Chapter 4 Glucocorticoids **37**
Antiinflammatory and Pulmonary Actions...............................37
Helping Athletes Recover Faster, Breathe Better,
and Burn More Fat 39
Endocrine and Immune Actions 39
Do They Work? Yes, No, and Maybe.......................................43
Pseudo-Cushing's Syndrome and Mineralocorticoid Actions 44
Conclusion..45
References..47

Chapter 5 Peptide–Protein Hormones................................49
Offsetting Endocrine Defects and Dysfunctions 49
Increasing Muscle Mass and Enhancing Oxygen Delivery 52
Ramping up Hormone Actions Within Cells.............................53
It Works, Within Limits 56
Acromegaly, Blood Hyperviscosity, and Death........................59
Conclusion..59
References..62

Chapter 6 Beta-2 Agonists...65
Asthma Prevention and Treatment...65
Breathing More and Building Muscle in Sport.........................67
Adrenergic and Anabolic Actions and Reactions 68
Asthmatics, Yes or Maybe! Everyone Else?.............................71
Tachycardia, Arrhythmias, Syncope..72
Conclusion..72
References..75

Chapter 7 Hormone and Metabolic Modulators................77
Androgen and Estrogen Modulation in Health and Disease 77
The Balance Between Gender-Based Hormones........................79
Selective Receptor Modulation Drugs—How They Work 81
Contradictions and Ambiguous Evidence.................................83
Thrombosis, Embolisms, Hot Flashes, and Hyperandrogenism 85
Conclusion..86
References..87

Chapter 8 Narcotics..91
In the Arms of Morpheus...91
The Athletic Need to Push Through the Pain............................92

The Opioid System and Pain 94
Substances That Work Too Well 96
Addiction, Addiction, and Addiction 97
Conclusion 98
References 100

Chapter 9 Beta Blockers 103
Cardiovascular Modulation and Health 103
Staying Calm, Cool, and Collected on the Battlefield of Sport 105
Blocking the Actions of Neurotransmitters and Hormones 106
Strong Evidence of Performance Enhancement 107
From Lethargy to Impotence: Undesirable Effects 108
Conclusion 109
References 110

**Chapter 10 Athlete Testing, Analytical Procedures, and Adverse
Analytical Findings 113**
WADA Testing Procedures 113
Biospecimen Analysis Techniques 115
Strengths and Weaknesses of the Bioanalytical Procedures 118
Ways in Which Athletes Avoid Detection 119
What Happens When an Athlete Gets a Positive Test Result? 121
Conclusion 124
References 126

Chapter 11 The Future of Performance Enhancement in Sport 129
Can Prevention and Detection Deter Doping? 130
Education of Athletes: Is That Enough? 130
Staying Ahead of the Doping Athlete: Is It Possible? 131
Genetic Doping: Is It the Future? 132
Should Doping be Legalized and Regulated? 133
Binary Gender Constructs and the Future 134
What's Next? 135
Conclusion 137
References 139

Index 141

Naama Constantini, MD
Director, Sport Medicine Unit, Shaare Zedek Medical Center, Jerusalem, Israel

Joseph W. Duke, PhD
Assistant Professor, Department of Biological Sciences, Northern Arizona University, Flagstaff, AZ, United States

Elizabeth S. Evans, PhD
Assistant Professor, Department of Physical Therapy Education, Elon University, Elon, NC, United States

Barnett S. Frank, PhD
Post-Doctoral Research Fellow, Sports Medicine Research Laboratory, University of North Carolina, Chapel Hill, NC, United States

Anthony C. Hackney, PhD, DSc
Professor, Schools of Public Health and Medicine, University of North Carolina, Chapel Hill, NC, United States

Hans Haverkamp, PhD
Associate Professor, Department of Environmental & Health Sciences, Johnson State College, Johnson, VT, United States

Martin Mooses, PhD
Vice-Director, Institute of Sport Sciences and Physiotherapy, Faculty of Medicine, University of Tartu, Tartu, Estonia

Kristin S. Ondrak, PhD
Associate Professor, American Public University, Charles Town, WV, United States

Maarit Valtonen, MD, PhD
Medical Director, Finnish Olympic Committee, High Performance Unit, Jyväskylä, Finland

FOREWORD

Doping in sport is nothing new—it has gone on since ancient times— and some athletes have and always will stoop to any means to win. Regrettably, in modern times, it seems the prevalence of athletes doping is on the rise; but the financial rewards with winning are astronomical and provide the motivation and temptation for many athletes to dope. The rewards may be great but the payback can be deadly as the use of many contemporary performance-enhancing drugs can result in serious morbidities and premature mortality.

I have known Dr. Anthony C. Hackney for a number of years. With over 200 peer reviewed research publications and a dozen book chapters, he is a prolific and a leading researcher in the sport and exercise sciences. In a 30-plus year career, his focus in the area of exercise endocrinology and specifically the development of dysfunction within the hypothalamic–pituitary–gonadal axis of athletes during exercise training has led to landmark findings.

Doping, Performance-Enhancing Drugs, and Hormones in Sport is Tony's fourth book project, two as an editor and two as a sole author. He and I worked together on the first editorial book project, *Endocrinology of Physical Activity and Sport*, which has been highly successful. That 2-year collaboration allowed me to work closely with him and see the depth of his knowledge, work ethic, focus and genuine passion for the topic. He has brought that same level of professionalism and commitment to this current book project.

I encourage you to read and enjoy this new book as it provides a concise synthesis and clear handling of an ever changing, complex and demanding scientific topic. This book is designed to provide the necessary background, understanding, and explanation for the student, the athlete, the coach and everyday individual who wants to fully comprehend the science of sports doping.

Naama Constantini, MD

International Olympic Committee Medical Commission, Working Groups on Female Athlete Health, Jerusalem, Israel

PREFACE

WHY ANOTHER BOOK ON DOPING?

I was asked by a colleague, "Why do we need another book on sports doping?" My answer was simple: "The books out there are good but very complex unless you already have a good understanding of biochemistry. They are mostly monolithic in length, and doping practices by athletes are constantly evolving at near light speed. I want to write something that is understandable, concise, and up to date." The response was "Good luck." That has been my objective in writing this book: Keep it understandable for laypersons as well as specialists, keep it accurate but manageable in the degree of technical detail, and try to capture the new developments in the prohibited and illegal practices found in sport doping as related to biochemistry and physiology.

DOES DOPING MATTER, SINCE EVERYONE IS DOING IT?

I am not an ethicist, but I know such people and have great respect for them. I will not venture into their field, because they are the experts and can address the ethics of doping far better than I (though in Chapter 11 I venture slightly into this area to discuss the implications of the legalization of doping). I am also guilty of a strong bias about doping, and bias is usually viewed as a fault in a scientist. My bias: *Doping in sport is wrong. It's flat-out cheating.* As a former athlete (track and field), I never enjoyed losing; but win or lose, I derived great pleasure from the competition. If I lost one day, that was because someone else was better on that day. There was always tomorrow and another chance to reverse the outcome. The underlying premise was always that you are each competing with your innate physical and mental talents, and the only issue is how hard you have worked in your training to develop those talents that make the difference. Doping destroys that premise and makes the playing field unfair. Some people in reading this are going to say that life isn't fair so why should sports be so? Again, I am no ethicist. I wish life were more fair. Because it is not, I feel that when we can make aspects of it as fair as possible, we should do so. To me, sport should be about fair play. It is an endeavor where lessons are learned that shape one's character and spirit in a positive way, and these lessons transcend sport and help us throughout life.

LANGUAGE: DOPING—WHAT'S IN A WORD?

The language used in this area of sports study is sometimes confusing for professionals and mystifying to the public. Terms range from highly technical to slangy: ergogenic aids, performance-enhancing drugs, dope or doping, roids, juice. All represent chemical compounds or techniques used by athletes to gain an unfair advantage. The book's title has three of these terms within it which are in some degree overlapping and redundant. That is intentional. I wanted the title to reach out to people of different backgrounds, knowledge levels, and experience who might not all recognize the slight differences and overlapping aspects of the terminology.

- *Doping* is the use of a banned chemical compounds (e.g., anabolic steroids) or techniques (e.g., blood doping) to improve sports performance. The substances and practices are called doping agents or just dope.
- *Performance-enhancing drugs* are substances banned by antidoping agencies. Some performance-enhancing substances, such as the diet supplement creatine, are not illegal or banned.
- *Hormones* are chemicals naturally produced in the body and serve to regulate such essential processes as digestion, metabolism, muscle growth, reproduction, and mood control. Artificially raising or lowing levels of hormones to enhance performance is doping.

IS THIS BOOK A GUIDE FOR THE ATHLETE WANTING TO DOPE?

The intent is emphatically not to provide guidance to athletes, coaches, bodybuilders, or anyone else on how to dope or avoid detection of doping. Its intent is deterrence. Scientific evidence is clear and unequivocal in showing that many doping agents can induce severe health consequences, possibly including premature mortality. If ethical, legal, and moral principles are insufficient deterrent, perhaps jeopardy to an athlete's quality and length of life will be. The intent is that, after reading this book, an athlete will understand how doping agents work, realize the great risk to their health and livelihood, and simply make the right decision and *not do it*.

HOW IS THE TEXT ORGANIZED?

Each of the chapters on specific drug categories (Chapters 2–9) is organized into six sections: history and medical use, basis for athlete use as a performance enhancer, mechanism of biochemical-physiological action, evidence of efficacy as a performance enhancer, side effects of use, and concluding key points.

ENJOY!

This work has been a labor of love. The physiology and biochemistry of how the human body works, and the bioanalytical chemistry aspects of understanding those workings, have always fascinated me and will continue to do so. The function of the body during exercise and sport is endlessly absorbing, and I am fortunate to be engaged with a profession about which I am passionate and constantly learning new and exciting things. I hope some of the passion and wonder of learning comes through to you the reader.

ACH

ACKNOWLEDGMENTS

I want to acknowledge the great help and support of the people at RTI International in North Carolina, especially Drs. Gerald T. Pollard and Brian F. Thomas, without whom this work would not be possible. My many thanks also to Dayle G. Johnson of RTI International for the cover design. Also, thanks to Katy Morrissey, Amy Clark, and Vijayaraj Purushothaman, who are respectively Acquisition Editor, Editorial Project Manager, and Production Project Manager at Elsevier.

I must recognize the constant encouragement, support, and cheer-leading of my wife Grace, who helped push and pull me through the challenging spells of bringing this work together.

Finally, my thanks to my doctoral student Amy Lane, who has been a devoted assistant, colleague, sounding-board, and supportive friend throughout this project.

My sincere thanks to all of you.

CHAPTER 1

Overview: Doping in Sport

Performance-enhancing substances, drug or otherwise, have been used in athletic competition since ancient times. The Greeks documented use more than two millennia ago. But if the phenomenon is nothing new, its prevalence and the magnitude of associated health risk have made it a major public issue. Hardly a week goes by without a news story about a major athlete caught doping or a medical finding about morbidities associated with pharmaceutical and nutritional supplements used by athletes. This chapter is an introductory overview of doping in sport—the essentials of what is happening, why it is happening, and what is being done about it.

HISTORY AND CURRENT STATE OF THE PROBLEM

The ancient Greeks used special diets and stimulating potions to fortify themselves in their Olympic Games. In the 19th and early 20th centuries, strychnine, caffeine, cocaine, and alcohol were commonly used, especially by distance runners and cyclists. An example of such early athletic use occurred in the 1904 St. Louis Olympic Games marathon. Thomas Hicks of the United States, the ultimate gold medalist, took raw eggs, injections of strychnine, and doses of brandy throughout the race. This was viewed as perfectly acceptable by the authorities, and many coaches and athletes developed their own concoctions and usage protocols.[1]

In the 1920s, sports governing bodies implemented restrictions on what came to be known as performance-enhancing drugs (PEDs) because of health concerns. However, these restrictions were essentially ineffective, because no testing was done to confirm or refute use. The pharmaceutical development of synthetic hormones such as testosterone in the following two decades made matters worse as these agents gained widespread use in sport. Public pressure on world sports authorities to introduce regular and rigorous testing was increased, in part due to the death from PED use of a Danish cyclist at the 1960 Rome Olympics and the media storm that followed. This very public

Doping, Performance-Enhancing Drugs, and Hormones in Sport. DOI: https://doi.org/10.1016/B978-0-12-813442-9.00001-8

event shocked the sporting world and lead to a public outcry. In 1967, the International Olympic Committee (IOC) took action and founded its Medical Commission, which established the first list of substances that athletes were prohibited from taking. Testing of athletes was finally introduced at the 1968 Winter Games in Grenoble and Summer Games in Mexico City that same year. Nonetheless, from the 1970s to the 1990s, PED use flourished as new designer drugs and techniques such as blood doping were devised by dishonest athletes and coaches. There were even large-scale state-sponsored doping programs. The former German Democratic Republic (East Germany) was a major culprit in such programs throughout its existence (1949–90), and recent evidence points to continued widespread state-sponsored doping by Russian sports agencies and the Russian government.[1,2]

Bioanalytical methods for detection of PEDs during the early period of test development lacked accuracy and reliability. This resulted in a lack of confidence in the results. In the 1990s, there was debate in the IOC, other international sports federations, and individual national governments about the exact definition of doping, enforcement policies, and appropriate punitive sanctions. Individual sanctions of athletes were often disputed and sometimes overruled in civil courts. As scandals mounted and disagreements persisted in sports organizations, the IOC convened the First World Conference on Doping in Sport in Lausanne, Switzerland in 1998. This landmark meeting resulted in establishment of the World Anti-Doping Agency (WADA) on November 10, 1999.[1]

WADA's objective is to promote and coordinate the fight against doping and PED use in all sports internationally, that is, to make the playing field level for all participants everywhere. Its success after nearly two decades is mixed. It has certainly not eliminated cheating with PEDs, but it has made great progress and continues to strive to keep sports clean of drugs. The mechanisms by which it polices sport and works with sports governing organizations are discussed in the following sections of this chapter (see also Chapter 11: The Future of Performance Enhancement in Sport).

PREVALENCE BY SPORT, COUNTRY, LEVEL OF COMPETITION

Accurate statistics on PED use are difficult to come by, not least because of the potential consequences of admission. Amateurs may be banned from competition. Professionals may be fined, banned, or entirely dismissed from the sport. But there are some indicative data on prevalence. One source indicates that 0.7% to 6% of male high-school athletes admit anabolic steroid use (a PED mimicking the male hormone testosterone), as do 0.2% to 5% of male college athletes and 9% of professional footballers. These numbers are tentative because self-reported; experts believe usage is higher, but athletes are fearful of admitting it. A thorough study by Uvacsek and associates found that 2% of elite athletes across all varieties of sports worldwide tested positive for a substance banned by WADA. These numbers are far lower than those for socially accepted recreational drugs. For example, 75% to 93% of male and 71% to 93% of female college athletes admit to using alcohol.[3] The PED numbers, while perhaps small, nonetheless point to a serious problem.

Table 1.1 shows data for doping cases in the Summer and Winter Olympics since 1968, the year testing procedures were implemented and records began to be officially kept by the IOC and WADA. As the number of tests administered increased over time, the number of positives showed no real trend. Table 1.2 breaks down the positives by sport and by country.[4] The data in the tables are for the Olympics, which represents a very limited sample of sporting events worldwide. While they suggest that a small percentage of athletes are engaged in doping, one should keep in mind that athletes and coaches know the Olympics are going to be the most scrutinized and tested sporting competition in the world, and they will take measures accordingly. This leaves open the question of how many Olympic athletes escaped detection and how many others are doping at the lower levels of competition where testing is limited or nonexistent.

WORLD ANTI-DOPING AGENCY AND LEGAL PROBLEMS

WADA works to implement and enforce the World Anti-Doping Code. The stated purposes of the code are

- To protect the Athletes' fundamental right to participate in doping-free sport and thus promote health, fairness and equality for Athletes worldwide and

Table 1.1 Doping Cases in Summer and Winter Olympic Games 1968–2014

Summer Olympics					Winter Olympics				
Year	Location	No. of Tests	No. Positive	% Positive	Year	Location	No. of Tests	No. Positive	% Positive
					2014	Sochi[a]	2473	10	0.40
2012	London	5051	9	0.18					
					2010	Vancouver	2149	3	0.14
2008	Beijing	4770	25	0.52					
					2006	Turin	1219	7	0.57
2004	Athens	3667	26	0.74					
					2002	Salt Lake City	700	7	1.00
2000	Sydney	2359	11	0.47					
					1998	Nagano	621	0	0.00
1996	Atlanta	1923	2	0.10					
					1994	Lillehammer	529	0	0.00
1992	Barcelona	1848	5	0.27	1992	Albertville	522	0	0.00
1988	Seoul	1598	10	0.63	1988	Calgary	492	1	0.20
1984	Los Angeles	1507	12	0.80	1984	Sarajevo	424	1	0.24
1980	Moscow	645	0	0.00	1980	Lake Placid	440	0	0.00
1976	Montreal	786	11	1.40	1976	Innsbruck	390	2	0.51
1972	Munich	2079	7	0.34	1972	Sapporo	211	1	0.47
1968	Mexico City	667	1	0.15	1968	Grenoble	86	0	0.00

[a]Currently, reanalysis of samples is occurring due to the Russian state-sponsored doping scandal [see Chapter 10 close-up]. Starting in 1994, the Winter Olympics was staggered by 2 years from the Summer Olympics.[1–5]

- To ensure harmonized, coordinated and effective antidoping programs at the international and national level with regard to detection, deterrence and prevention of doping.[5,6]

The detection process is focused on ensuring that athletes are not using any of the substances on the WADA Prohibited List, which is extensive and regularly updated. Urine and/or blood specimens are collected at in-competition and out-of-competition settings by a WADA representative and are brought to a WADA accredited laboratory, of which there were 34 as of 2017, for analysis by state-of-the-art bioanalytical chemistry techniques. WADA has strict policies and procedures on handling to guard against contamination or falsification—that is, a chain of custody protocol. Results are forwarded to the sports association of the athlete. The athlete is then notified of the test results and of

Table 1.2 Positive Doping Cases in Olympic Games 1968–2014 by Sport and Country[1–5]			
By Sport		**By Country**	
36	Weightlifting	10	Austria
28	Track & field	9	Greece Russia
12	Cross country skiing	8	USA
8	Equestrian	7	Bulgaria Hungary
6	Ice hockey Wrestling	5	Poland Spain
5	Cycling	4	Germany Sweden
3	Biathlon Modern pentathlon Volleyball	3	Great Britain Mongolia Norway Ukraine
2	Baseball Gymnastics Judo Rowing Swimming Shooting	2	Belarus Brazil Canada Czechoslovakia India Ireland Italy Japan Puerto Rico Romania
1	Alpine skiing Basketball Boxing Canoeing Sailing	1	Afghanistan, Algeria, Armenia, Australia, Bahrain, China, Croatia, Finland, Iceland, Iran, Kenya, Latvia, Lebanon, Lithuania, Moldova, Monaco, Morocco, Myanmar, Netherlands, North Korea, Slovakia, Turkey, Unified Team (CIS), USSR, Uzbekistan, Vietnam, West Germany

any sanctions if a banned substance is found (see Chapter 10: Athlete Testing, Analytical Procedures, and Adverse Analytical Findings).[5]

WADA has no legal mechanism itself by which to impose sanctions on athletes. That is up to other national and international agencies and organization such as the United States Anti-Doping Agency, the International Association of Athletics Federation, and the IOC. Sanctions range in severity from public warning to multiyear suspension from all competitions. A common sanction is "loss of results," which means voiding the athlete's specific performance results, entire record, or standing in competitions. In many cases, an athlete who has won a medal or prize and subsequently tested positive is required to

Figure 1.1 Turkey's Asli Cakir Alptekin (R) wins gold ahead of her compatriot second placed Gamze Bulut (3rd L) and third placed Bahrain's Maryam Yusuf Jamal (C) in the women's 1500-m final during the London 2012 Olympic Games at the Olympic Stadium August 10, 2012. REUTERS/Lucy Nicholson.[7]

surrender the medal or prize, and the victory is declared null and void. In the most serious cases, athletes may be barred for life from sport competitions (see Chapter 10). A memorable example is Canadian sprinter Ben Johnson, who set a world record and won the gold medal in the 100-m men's final at the 1988 Seoul games and was later disqualified for using a PED (anabolic steroids) and his record and medal were taken away. This race has been referred to in sports literature as the "dirtiest Olympic final." The distinction was bestowed because six of the eight finalists had confirmed doping allegations or sanctions against them during their sports careers. Such occurrences are not a thing of the past; a new candidate for the dirtiest Olympic final is the women's 1500-m event at the London 2012 games. Five of the first seven women to cross the finish line (identified in Fig. 1.1) ultimately failed a doping test.[7]

ETHICAL ISSUES OF USAGE, SPORTSMANSHIP, CHARACTER

Why would a coach or athlete consider using a PED, in light of the potential health, social, financial, and legal consequences? There are three common explanations:

The Means-To-An-End Philosophy

Winning isn't everything. It's the only thing.

Vince Lombardi

This ethos permeates sport. Put simply, winning is so important that cheating becomes an adjunct to physical training, consequences be damned.

Rewards are Too Tempting

The potential to make a career and a living from participation in sport has existed for nearly two centuries. Financial rewards have reached exorbitant heights in the professional sports ranks such as the National Football League, the National Basketball Association, and Major League Baseball. For example, in 2016 baseball pitcher Clayton Kershaw's annual salary was $32.8 million, and in 2017 basketball player Stephen Curry negotiated a $34 million salary. Even sports once considered bastions of amateurism have become professional and lucrative. In the 14-plus annual competitions of the Diamond League, which is part of the International Association of Athletics Federations (the track and field world governing body), the athletes are competing for a total purse of US$450,000. National sports agencies such as the United States Olympic Committee pay athletes $25,000 for gold, $15,000 for silver, and $10,000 for bronze medals at the Olympic Games. Product endorsement can be extremely lucrative. Olympic sprint champion Usain Bolt earned an estimated $32.5 million between June 2015 and June 2016 from product endorsement contracts.[8]

What's So Wrong if Everyone's Doing It?

The battle of Waterloo was won on the playing fields of Eton.

Duke of Wellington (apocryphal)

The public schools of England saw team sports as a way to teach boys discipline, the importance of hierarchy, the skills, the codes of honor, and the leadership qualities necessary to run the Empire. Actual or apocryphal, the Iron Duke's statement rings true, that sports and gaming activities teach aspects of behavior, social justice, rules, values, and acceptability, as well as improving physical health. Doping shatters those social constructs and creates an alternative perspective on what is right, wrong, and acceptable. Not all people play sports, but those who do have the opportunity to live out ethical choices that carry over into their own lives and the lives of those spectators for whom they are role models. Adolf Ogi, special adviser to the United Nations Secretary-General, put it this way: "Sports represents the best school of life by teaching young persons the skills and values they need to be good citizens."[9]

Research evidence bears out this line of thinking and supports the proposition that sports participation in youth can positively affect aspects of personal development such as self-esteem, goal-setting, and leadership development.[10] For example, Goldberg and associates examined over 3200 high-school athletes and reported more positive behavioral and social attitudes within their sample than in nonathletes; these sportsmen also had a reduced likelihood of participating in illegal acts.[11] The US Department of Education in a longitudinal study of nearly 10,000 high schoolers found that after graduation athletes were more likely than nonathletes to attend college and get advanced degrees; the sports team captains and most-valuable players had these achievements at even higher rates.[12] The benefits extend to the workplace. A survey of 400 female corporate executives found that 94% played a sport, and 61% said that it contributed to their career success and development of interpersonal skills.[13]

ARE THERE SITUATIONS IN WHICH ATHLETES CAN USE PEDS?—THERAPEUTIC USE EXEMPTION

Like everyone else, athletes have health problems, some of which require pharmaceutical intervention. Healthcare providers might need to prescribe as part of the "standard of care" of a medical condition a drug that is on the WADA list of banned agents. In fact, many PEDs have legitimate medical uses, as discussed in later chapters. In this situation, WADA permits athletes to apply for a therapeutic use exemption (TUE). In 2014, 897 TUEs were granted, a 41% increase from 2013.[14]

However, WADA has strict guidelines for TUEs issuance. For example, testosterone replacement therapy may be granted to only those male athletes who have an organic androgen deficiency.[14,15] Deficiency here is divided into primary medical conditions (e.g., Klinefelter syndrome, cryptorchidism, direct testicular trauma and torsion, radiation, chemotherapy) and secondary medical conditions (e.g., pituitary disorders, sickle-cell disease, anatomical issues, androgen deficiency).[15] These are separate and distinct from the functional androgen deficiencies resulting from severe stress, overtraining, or chronic systemic illnesses, where a TUE would not be granted. The WADA guidelines state, "TUE should only be approved for androgen deficiency that has an organic etiology," not for a functional disorder and not in females.[15]

The TUE application must be submitted to WADA with information on the medical history, physical examination, and laboratory evaluation by the primary physician, commonly an endocrinologist. Furthermore, the dose of the PED administered, if approved, must not allow the athlete an unfair advantage.[16] Some sports impose their own additional regulations, dictated by individual governing bodies, which differ slightly from the WADA TUE guidelines. Hence athletes and their physicians must take great care to ensure they are following the steps and regulations before medical treatment with a drug that could be viewed as a PED.

It is also important to recognize that physician have taken the Hippocratic Oath: First, do no harm. This can result in medical situations where the standard of care for treating an athlete may require the use of a banned substance, and the schedule of training and competition does not allow enough time for a TUE application (see the close-up in Chapter 8: Narcotics). Here the necessity to provide appropriate and ethical care can outweigh the doping consideration. The physician must document the rationale and treatment in detail and provide the record to the national antidoping agency and WADA.[14,15]

CONCLUSION

The establishment of WADA in 1999 was a signal event in the decades-long effort to monitor and work toward the elimination of PED use in sport. Testing techniques have improved and consciousness has been raised about the negative health and ethical effects. But doping is entrenched, the task is demanding, and WADA is still a work in progress trying to stay one step ahead of the cheaters. The following chapters deal with the complexity of this challenging problem.

Close-Up: Sports Doping in Ancient Civilization—Nothing New Under the Sun

The use of drugs, plant extracts, and diets to enhance performance has been a feature of competition since the beginning of recorded history.[1,17] As early as 1400 BC, the Susruta of India advocated the ingestion of animal testis tissue to cure impotence, and the ancient Egyptians also accorded medicinal powers to the testicles and prescribe their use for individuals.[17] The Greek physician Aretaeus the Cappadocian (c. AD 150) suggested the endocrine function of the testis, in particular the anabolic and androgenic effects of using extracts as a supplement. There are reports of warriors in many cultures eating the heart of an animal to promote bravery and the brain to improve intelligence, and of cannibalizing defeated opponents.

Greek Olympic athletes also sought to improve performance with devices (Fig. 1.2) and special diets.[18] Charmis, the Spartan winner of the *stade* (183 m) in the Games of 668 BC, reportedly ate only dried figs. Other athletes of this time were advocates of wet cheese and wheat meal diets. Dromeus from Stymphalos, who won the *dolichos* (between 1.6 and 4.8 km) twice at Olympia, twice at Delphi, three times at Isthmia, and five times at Nemea, followed a primarily meat diet.[1,17,18] PEDs also were used, in the form of brandy and wine concoctions, hallucinogenic mushrooms, and sesame seeds.[19] The obsession with winning at all cost seems to be not so modern a concept. These practices were all part of Greek culture and were viewed as perfectly acceptable. The purpose of the Games was to show off the athletic prowess of the various city states, but in reality they served as a surrogate to signal military preparedness and readiness to fight, so winning at all cost was critical to make a political statement (Fig. 1.2).[1]

*Figure 1.2 Ancient Greek long jumpers using hand weights (*halteres*) to propel themselves further.*

The steps taken by athletes to best each other were extensive, but actual cheating was frowned upon and punished by fines, public flogging, and statewide banning from further competition.[20] The 2nd-century AD traveler Pausanias wrote, "It is the custom for athletes, their fathers and their brothers, as well as their trainers, to swear an oath upon slices of boar's flesh that in nothing will they sin against the Olympic Games. The athletes take this further oath also, that for 10 successive months they have strictly followed the regulations for training."[21]

Doping control as we think of it was not enforced in ancient times, but honor was expected and dishonest means of winning were not acceptable or condoned.

REFERENCES

1. Yesalis CE, Bahrke MS. History of doping in sport. *Int J Sports Stud*. 2002;21(1):42−76.

2. McLaren Report. Independent person WADA investigation of the Sochi allegations. December 9, 2016. <https://www.wada-ama.org/sites/default/files/resources/files/mclaren_report_part_ii_2.pdf>; 2016 (accessed 03.08.2017).

3. Uvacsek M, Nepusz T, Naughton DP, et al. Self-admitted behavior and perceived use of performance-enhancing vs psychoactive drugs among competitive athletes. *Scand J Med Sci Sports*. 2011;21:224−234.

4. Reardon CL, Creado S. Drug abuse in athletes. *Subst Abuse Rehabil*. 2014;5:95−105.

5. World Anti-Doping Agency. Anti-Doping Code 2015. <http://www.usada.org/wp-content/uploads/wada-2015-world-anti-doping-code.pdf>; 2015 (accessed 03.07.2017).

6. Tandon S, Bowers LD, Fedoruk MN. Treating the elite athlete: anti-doping information for the health professional. *Missouri Med*. 2015;112(2):122−128.

7. Phillips M. London 1500 stakes claim as 'dirtiest race'. *Reuters*. July 20, 2016. <http://www.reuters.com/article/us-olympics-rio-athletics-1500-idUSKCN100331>; 2016 (accessed 2808.2017).

8. Forbes Magazine. The worlds 100 highest paid athletes: by the numbers. June 8, 2016. <https://www.forbes.com/sites/kurtbadenhausen/2016/06/08/the-worlds-100-highest-paid-athletes-2016-by-the-numbers/#569344c557eb>; 2016 (accessed 03.18.2017).

9. United Nation Office on Drugs and Crime. Everyone Wins! Global Youth Network Team—Prevention, Treatment and Rehabilitation Unit. Vienna, Austria; 2004.

10. Government Accountability Office. K-12 education: school-based physical education and sports programs. GAO Office of Public Affairs. GAO Report # 12 350. February 29, 2012.

11. Goldberg L, MacKinnon DP, Elliot DL, et al. The adolescents training and learning to avoid steroids program: preventing drug use and promoting health behaviors. *Arch Pediatr Adolesc Med*. 2000;154(4):332−338.

12. U.S. Department for Education. National Center for Education Statistics. What is the status of high school athletes 8 years after their senior year? NCES 2005-303; September 2005.

13. EY Women Athletes BusinessNetwork: Report 2014. <http://www.ey.com/Publication/vwLUAssets/EY-where-will-you-find-your-next-leader/$FILE/where-will-you-find-your-next-leader-report-from-EY-and-espnw.pdf>; 2014 (accessed 03.06.2017).

14. WADA Annual Report, 2014. <https://www.wada-ama.org/en/>; 2014 (accessed 02.01.2017).

15. World Anti-Doping Agency. Androgen deficiency/male hypogonadism: TUE Physician Guidelines. WADA Headquarters, Montreal (Quebec) H4Z 1B7, Canada; 2015.

16. Handelsman DJ, Heather A. Androgen abuse in sports. *Asian J Androl*. 2008;10(3):403–415.

17. Hoberman J, Yesalis C. The history of synthetic testosterone. *Sci Am*. 1995;272(2):76–81.

18. Finley M, Plecket H. *The Olympic Games: The First Thousand Years*. London: Chatto & Windus; 1976.

19. Voy R. *Drugs, Sport, and Politics*. Champaign, IL: Leisure Press; 1991.

20. Smithsonian.com. The ancient history of cheating in the Olympics. August 3, 2016. <http://www.smithsonianmag.com/history/ancient-history-cheating-olympics-180960003/>;   2016 (accessed 03.06.2017).

21. Jones WHS, Omerod HA. *Pausanias—Description of Greece*. Loeb Classical Library Volumes. Cambridge, MA: Harvard University Press; London, William Heinemann Ltd; 1918.

CHAPTER 2

Anabolic Androgenic Steroids

Perhaps, the best-known class of performance-enhancing drugs (PEDs) in the sporting world is anabolic androgenic steroids, also called anabolic steroids, steroids, or just "roids." These are chemical agents that mimic the actions of the male sex hormone testosterone, which is associated with anabolic physiological actions (i.e., synthesis or building up) such as increased muscle mass and strength. They have played a major role in the doping scandals of sport and continue to be extremely popular despite their severe health consequences.

TREATMENT OF ENDOCRINE DISRUPTIONS

Although its biological effects have been known since antiquity, the name testosterone was coined in 1935, when Ernest Laqueur isolated it from bull testes.[1] The prior scientific steps to this isolation were long, as John Hunter had transplanted testes into capons in 1786 and Adolph Berthold had postulated internal secretion from his testicular transplantation experiments in 1849. In the 1900s, monkey testicular preparations were used for therapy, popularized by the self-experiments of French physician Charles Edouard Brown-Séquard. In the 1920s, Sergio Voronoff transplanted testes from animals into men, but the effectiveness was questioned. Modern androgen therapy began when testosterone was chemically synthesized independently in 1935 by the German Adolf Butenandt and the Croatian-Swiss Leopold Ruzicka, both Nobel Laureates.[2]

In early experiments, testosterone was found ineffective if administered orally. However, if pharmaceutically modified into a different chemical structure, 17α-methyl testosterone, dubbed an anabolic androgenic steroid, it could be orally absorbed, although this version had deleterious side effects. Anecdotal evidence that steroids enhanced physical strength and aggression led to experiments on soldiers by Germany during the Second World War, but there are no conclusive reported findings from these experiments in the literature.[3] In the

Doping, Performance-Enhancing Drugs, and Hormones in Sport. DOI: https://doi.org/10.1016/B978-0-12-813442-9.00002-X

1950s, longer-acting injectable testosterone enanthate became the preferred therapeutic modality. In the late 1950s and into the 1960s, researchers concentrated on the chemical modification of steroid drugs to emphasize anabolic effects over androgenic effects. This period coincided with the expanded use of testosterone and its steroid drug derivatives in sports doping. In the 1970s, the orally effective testosterone undecanoate was added to the evolving spectrum of pharmaceutical preparations used by athletes. In the 1990s, there were calls within the medical endocrinology community for testosterone preparations that would produce physiological serum levels without the serious side effects of injected versions, which lead to the development of the first transdermal scrotal film administration. Nonscrotal skin patches followed shortly thereafter, and by the turn of the century transdermal testosterone gels were in use.[4,5] Short-acting buccal testosterone and long-acting improved versions of injectable testosterone undecanoate are now available.[2,4,5]

Anabolic steroids are used medically for several conditions, such as cachexia (muscle wasting) associated with AIDS, certain cancers, severe burns, and renal failure, where nutrition intervention and standard care are ineffective. Use for age-related sarcopenia in men and women has been advocated.[4] This approach focuses on the potential anticatabolic aspects of testosterone derivatives to lessen proteolysis and enhance proteogenesis, principally in skeletal muscle.[6]

For hormone replacement therapy, testosterone preparations are effective in male hypogonadism as well as for male hormonal contraception (i.e., where progestogens are administered to inhibit gonadotropin secretion). Steroids also stimulate erythrocyte synthesis, which can be used in the treatment of hypoplastic anemia; the recent availability of recombinant human erythropoietin and its analogues has decreased this usage. In the past, postmenopausal women were given steroids for osteoporosis, but this is no longer advocated because of the success of estrogen replacement therapy and the introduction of bisphosphonates to aid in accretion of bone mineral content.[4]

THE ATHLETIC DRIVE TO INCREASE MUSCLE MASS

Athletes use steroids to facilitate physiological development beyond that achieved with exercise training alone. They are attempting to increase body weight, strength, power, speed, or endurance.[7,8] When used properly in combination with correct exercise training, these pharmaceutical agents are highly effective in both men and women.[8,9] Table 2.1 shows some of the most popular agents. Usage occurs in many sports and is quite prevalent in athletics (track and field, mostly throwing events), weightlifting, bodybuilding, and American football. Regrettably, usage is rising in nonathletes to enhance body image and, to some degree, muscular function.[10]

Prevalence of Usage

Because steroids are banned by antidoping agencies and in many situations illegal, prevalence is difficult to determine, because few users are willing to admit it. In self-report studies, estimates range from 5% to 30% of the professional and amateur athlete populations, particularly in sports were body size, muscular strength, and power are criterial, even for very young athletes.[11–14] Buckley and associates studied a North American sample of 3400 high school seniors and found that over 6% used steroids periodically; most users were athletes, but over a

Table 2.1 Anabolic Steroids Popular among Strength-Power Athletes and Body Builders[11]	
Anabolic Steroid Agent	**Brand or Popular Name**
nandrolone	Deca-Durabolin
methandrostenolone	Dianabol, Averbol, Danabol, Metanabol, Naposim, Vetanabol
oxandrolone	Oxandrin, Anavar, Oxandrin, Lonavar
oxymetholone	Anadrol, Anapolon
trenbolone	Finaplix
boldenone	Equipoise (veterinary)
tetrahydrogestrinone	THG—"The Clear"
norbolethone	Genabol
desoxymethyltestosterone	Madol, Pheraplex
danazol	Danatrol, Danoval
fluoxymesterone	Halotestin, Ultandren
methyltestosterone	Methitest, Agovirin, Android, Metandren, Oreton, Testred, Virilon
stanozolol	Winstrol, Stromba
testosterone cypionate	Depotestosterone
testosterone enanthate	Delatestryl, Testostroval, Testro LA, Andro LA, Durathate, Everone, Testrin, Andropository
testosterone propionate	Agrovirin, Andronate, Andrusol-P, Masenate, Neo-Hombreol, Oreton, Perandren, Synandrol, Testoviron

third were nonathletes who cited improved body appearance as their main objective.[14] Twenty percent of the users had obtained their drugs from a medical professional. Research by Yesalis et al. in the 1990s put the total number of users in the United States at over 1 million.[15] There is no evidence that the prevalence has declined, in spite of the warnings by health care providers of the dangers.[13–15] Recent work by Albertson et al. suggests that overall usage, especially among recreational athletes, may be increasing in the United States and worldwide.[16]

Banned Substances

After the 1972 Munich Olympics, many athletes admitted having used steroids as part of their training preparation.[9] The International Olympic Committee (IOC) had already formally banned certain drugs and practices at this point, but the IOC did not ban anabolic steroids until 1975. This was principally because accurate and reliable bioanalytical detection techniques did not exist. These revelations prompted the IOC to encourage the development of better methods. In 1974, when new methods became available, the IOC added steroids to its banned list. Metabolites in urine were detectable by gas chromatography-mass spectrometry (see Chapter 10: Athlete Testing, Analytical Procedures, and Adverse Analytical Findings).[7] In addition to the IOC and the World Anti-Doping Agency, nearly all antidoping agencies and major sports-governing organizations in the world have banned the use of steroids that are taken for the purpose of unfairly and artificially improving performance in competition.

INTRACELLULAR BIOCHEMICAL AND NUCLEAR ACTIONS

Anabolic steroids affect both physiological and psychological function and are mimics of testosterone (see Chapter 7: Hormones and Metabolic Modulators, Fig. 7.2).[8] Proposed mechanisms are (1) activation of steroid hormone receptors in skeletal muscle cells, (2) an anticatabolic effect in skeletal muscle, and (3) motivational psychological effects. The extent to which each of these mechanisms proportionally contributes to the overall physiological adaptation and performance change is uncertain. Most likely, the elements function synergistically.

Steroid Hormone Receptors

Androgenic hormones and anabolic steroids activate androgen receptors in skeletal muscle fibers, which stimulate the promoters of specific genes and induces proteogenesis (protein synthesis) via the nuclear mechanisms of transcription and translation.[9,17] Specifically, proteins such as those associated with the muscular contractile process (actin and myosin) are increased in density. This allows the muscle to have greater focused output. A key protein also expressed in this process is the hormone insulin-like growth factor-1, which directly promotes muscle growth and is strategic in androgen receptor signaling.[18,19] Receptor signaling is critical in that it amplifies the physiological effect of steroids (and of endogenous testosterone).

Several transcription factors up-regulate androgen receptor function.[17] Of these, supervillin, a 205-kDa actin-binding protein, is key because it is regulated by anabolic-androgen-based hormones; steroids are proposed to work by the same mechanism.[20] Put simply, testosterone and anabolic steroids activate androgen receptors, supervillin promotes more receptor expression, and testosterone regulates the expression of supervillin, a classic endocrine amplification effect.

For a steroid to have maximal effect on performance, exercise training seems to be necessary. Several research groups have shown greater improvement in experienced weightlifters than in novices.[7,9] Experienced lifters were enabled to train with heavier weights and so produce greater muscle tension. Steroid effectiveness also appears dependent upon the number of unbound androgen receptor sites in skeletal muscle, which are increased by intensive resistance training.[9,17,21]

Anticatabolic Actions

The anabolic effects of testosterone and steroids are mediated by an antiglucocorticoid action, principally on cortisol. Cortisol is catabolic in that one of its actions is to induce proteolysis.[22] Cortisol secretion is highly stimulated by exercise, which activates the processes (e.g., gluconeogenesis) in which amino acids from proteins are used for energy metabolism. This in turn depletes the free amino acid pool of proteogenesis precursors.[9,22] Glucocorticoid receptors are found in many tissues, including skeletal muscle fibers. Testosterone and steroids bind to these receptors to block the binding of cortisol and prevent the induction of muscle protein breakdown.[7,17,23,24] This effect of blocking receptors could enhance recovery from exercise training by allowing amino acid precursors to be available for tissue repair and regeneration, although this is a point of debate in the scientific community. Athletes say that steroids help them train more intensely and recover faster, but it is impossible to determine how much of this effect is physiological versus psychological.[7,10]

Psychological Effects

An important aspect of steroid efficacy may be enhancement of the motivational state.[12] This notion is supported by athletes' reports of enhanced emotional wellbeing, euphoria, aggressiveness, and exercise stress tolerance, all positive motivators that allow the athlete to train harder. This in turn allows a greater training stimulus to be presented to muscle fibers. The result is improved physical adaptation for performance.[24] Also, one cannot discount the possibility of a powerful placebo effect in these factors.[12]

EARLY PARADOXES AND LATER CONFIRMATIONS IN RESEARCH FINDINGS

Early research suggested equivocal effects of steroids on performance,[7] but recent evidence has clearly documented increased muscle mass and maximal voluntary muscle strength enhancement.[9,25,26] For ethical reasons, most of the early studies did not test the high dose levels used by self-administering athletes; this could account for the disagreement with anecdotal reports of substantial effects. Though the anecdotal reports were tainted by lack of controls for confounding factors such as extent of training and the placebo effect, they suggested body mass gain of 10 to 20 kg and up to 30% increase in strength.[7,9] These very large effects are not typically seen in individuals with regular training regimens of comparative time periods when no PEDs are use.[24] Several well-controlled studies have found smaller but significant increases in total body mass, lean body mass, and muscular strength, and concurrent reductions in fat mass.[9,24] In most of these studies, however, because of health concerns, the duration of steroid administration was shorter than used by most self-administering athletes.

Most studies of endurance athletes—distance runners and triathletes—have shown no substantial benefit from steroids on maximal oxygen consumption, a key determinate of aerobic endurance capacity.[7,9,24] It appears that endurance athletes use steroids to speed recovery from training bouts. This seems logical, since more endurance training improves cardiovascular-respiratory function and thereby aerobic performance, and steroids should offset the decline in testosterone that normally follows high volume exercise.[23] Steroids also slightly promote erythropoiesis, increasing the oxygen carrying capacity of the blood, which is highly beneficial to endurance athletes, but there are far more beneficial PEDs than steroids for promoting erythropoiesis (see Chapter 5).[27,28]

EFFECTS OF ABUSE, REVERSIBLE, AND NONREVERSIBLE

Steroids are harmful to several physiological systems, with severity ranging from mild to serious. Some effects go away when use is stopped, others are permanent, and a few are life-threatening. Table 2.2 shows the major side effects.[11]

Table 2.2 Side Effects of Anabolic Steroid Use by Tissue or Organ System[11]	
Tissue or Organ	**Effect**
Skin	
	Severe acne on face and back Baldness Stretch marks Bloating Skin infection at injection sites
Muscle and joint	
	Aching joints Muscle cramps Tendon rupture
Heart	
	High blood pressure High cholesterol Heart disease Heart attack
Liver	
	Transient serum enzyme elevations Acute cholestatic syndrome Chronic vascular injury (peliosis hepatis) Hepatic tumors (adenomas and hepatocellular carcinoma)
Brain	
	Headache Stroke
Gastrointestinal	
	Nausea Vomiting Diarrhea
Bone	
	Epiphyseal plate closure

A very important and dangerous side effect is the increased risk of cardiovascular disease. Blood total cholesterol levels increase and high-density lipoprotein cholesterol (HDL) declines markedly, especially with oral administration.[29] It appears that steroids influence the activity of the hepatic enzymes triglyceride lipase (HTL) and lipoprotein lipase (LPL). HTL is primarily responsible for clearing HDL, while LPL facilitates the cellular uptake of free fatty acids and glycerol. Steroids (and androgenic hormones) stimulate HTL activity, reducing serum HDL.[29,30] Steroids may also provoke a transient hypertensive state; high doses increase diastolic blood pressure.[29–32] Steroids also produce structural changes in the heart that decrease myocardial ischemic tolerance.[29,33] Echocardiographic studies in bodybuilders using steroids showed mild after-load hypertrophy of the left ventricle with decreased diastolic relaxation, resulting in decreased diastolic filling and hence compromising the pumping capacity of the heart.[32] Some investigators have associated development of cardiomyopathy, myocardial infarction, and cerebrovascular accidents with chronic steroid use, but not all of the evidence is conclusive.[34,35] Such physiological effects have negative influence on overall cardiovascular health risk factors. For this reason, ethical considerations prevent rigorous scientific study of long-term steroid use, but case studies of self-administering users support the association.[36]

Many other serious side effects can damage overall health and increase morbidity and mortality.[29,36] Hepatotoxicity, fatty liver, and liver neoplasm are especially noteworthy, and there are anecdotal reports of permanent hypothalamic–pituitary–testicular suppression and resultant reproductive dysfunction.[36]

Several side effects are to some degree gender-specific. Men have shrinkage of the testicles, decreased endogenous testosterone production, decreased sperm count leading to reduced fertility, and enlarged breasts (gynecomastia). Changes in women revolve around androgenic effects: development of facial hair, a deeper voice, oligomenorrhea or amenorrhea leading to reduced fertility, and reduction in breast size.[11,12] Psychological effects are frequently reported: "roid rage" (highly aggressive and violent behavior), increased anger or irritability, severe mood swings, paranoia, anxiety, panic attacks, and depression.[11]

OTHER ANABOLIC SUBSTANCES

An agent that has received much press coverage is androstenedione, a precursor for both androgen (testosterone) and estrogen (estradiol) based hormones. It is a weak androgen and is poorly converted to more potent forms. When orally administered, most of it is metabolized to testosterone glucuronide and other metabolites before release into the general circulation.[37,38] Several well-controlled studies of its effects have been done.[9,39] The results, based upon evaluation of circulating androgen levels and muscle strength or morphology, failed to support a substantial enhanced anabolic effect.

Dehydroepiandrosterone (DHEA) and its sulfate (DHEA-S) are the most abundant circulating steroid hormones in humans. DHEA, like androstenedione, may serve as a precursor for androgens and estrogens, and direct action through nuclear and G-protein receptors have been suggested by some investigators.[40] Theoretically, use of DHEA or DHEA-S as a supplement could have beneficial anabolic effects. In older adults, circulating androgen levels and subjective ratings of physical and psychological well-being have been noted.[41] The effects are perhaps strongest in patients with adrenal insufficiency and those on high dosage glucocorticoid therapy.[9] Effects on muscle strength and performance in athletes remain questionable but physiologically plausible.[9,39]

CONCLUSION

Anabolic androgenic steroids are the best-known class of PEDs in the sporting world. They are powerful mimics of the male sex hormone testosterone, and hence have major anabolic physiological actions such as increased muscle mass and strength. Effectiveness as PEDs comes at a cost to the athlete, as there are serious side effects which can lead to many comorbidities and potentially premature mortality.

Close-Up: Victims of the East German Medal Machine 1970s to 1990s

The German Democratic Republic (GDR; East Germany) came to an end in 1990 shortly after the fall of the Berlin wall. The government had conducted a decades-long program of forced administration and distribution of PEDs to many of its elite athletes. This was done to bolster the communist state's international image and prestige by winning medals in international championships, especially the Olympics. The program was officially known as State Plan 14.25 and began around 1969. It was highly successful. In Olympic competition alone from 1956 to 1988, the medal count amounted to 203 gold, 192 silver, and 177 bronze, remarkable for a country of less than 20 million people.

These drugs were given either with or without the knowledge of the athletes. Many of them now suffer from PED-related health problems. Weightlifter Roland Schmidt received massive amounts of anabolic steroids and growth hormone. Ultimately, he grew size 36DD breasts and had to have them surgically removed, as his body had stopped producing testosterone. Female shot putter Heidi Krieger suffered the opposite effect. Because of chronic testosterone and steroid use over years, she essentially lost femininity. Eventually, she changed sex and now lives as a man.

Ines Geipel was part of a 1980s world-beating East German sprint team (track and field). She is now a college lecturer and has set up a support group for women who were GDR victims of doping. She estimates that 10,000 to 15,000 athletes were part of the doping program. Children as young as eight were supposedly being doped and used as guinea pigs and are now suffering severe health consequences. Said Geipel,

The bodies are broken, and so are the souls. Naturally there was great gynaecological damage because the women were taking men's hormones. We have seen still births, infertility, and disabled children born to former athletes.[42]

As I wrote this piece, I realized that I am of an age such that I competed against some of these GDR athletes. At the time and for years afterward, I only viewed them as cheaters. Now, as their stories have come out, I grasp that they were pawns in a geopolitical game. Many of them did not know they were being given PEDs and thought they were receiving nothing more than vitamins. And the ones who did know were in an unbelievably difficult situation as they were living in a totalitarian state. Refusing might have placed their lives and livelihoods, and that of their families, in jeopardy. Now I have a tremendous amount of sympathy and empathy for these athletes. They were victims of the times.

REFERENCES

1. Knegtmans PJ. Ernst Laqueur (1880–1947): the career of an outsider. In: Maas A, Schatz H, eds. *Chapter 7 of Physics as a Calling: Studies in Honor of A.J. Kox.* Leiden: Leiden University Press (LUP); 2013.

2. Nieschlag E. The history of testosterone. *Endocr Abstr.* 2005;10:S2.

3. Yesalis CE, Bahrke MS. History of doping in sport. *Int J Sports Stud.* 2002;21(1):42−76.

4. Kicman AT. Pharmacology of anabolic steroids. *Br J Pharmacol.* 2008;154(3):502−521.

5. Nieschlag E. Testosterone treatment comes of age: new options for hypogonadal men. *Clin Endocrinol.* 2006;65(3):275−282.

6. Hackney AC, Anderson T, Dobridge J. Exercise and male hypogonadism: testosterone, the hypothalamic−pituitary−testicular axis and exercise training. In: Winters SJ, Huhtaniemi IP, eds. *Male Hypogonadism: Basic, Clinical and Therapeutic Principles.* New York: Springer − Humana Press; 2017:257−280.

7. American College of Sports Medicine. *Position Statement on the Use of Anabolic Steroids.* Indianapolis: ACSM Publishing; 1990.

8. Rogol AD, Yesalis CE. Clinical review—anabolic-androgenic steroids and athletes: what are the issues? *J Clin Endocrinol Metab.* 1992;74:465−469.

9. Rogol A. Sex steroid and growth hormone supplementation to enhance performance in adolescent athletes. *Curr Opin Pediatr.* 2000;12:382−387.

10. Sagoe D, Molde H, Andreassen CS, et al. The global epidemiology of anabolic-androgenic steroid use: a meta-analysis and meta-regression analysis. *Ann Epidemiol.* 2014;24 (5):383−398.

11. Drug record − Anabolic steroids. National Institute of Diabetes and Digestive and Kidney Diseases. <https://livertox.nih.gov/AndrogenicSteroids.htm>; 2017 (accessed 02.05.07).

12. Mottram DR, George AJ. Anabolic steroids. *Baillieres Best Pract Res Clin Endocrinol Metab.* 2000;14:55−69.

13. Laure P. Epidemiological approach of doping in sport. *J Sports Med Phys Fitness.* 1997;37:218−224.

14. Buckley WE, Yasalis CE, Friedl KE, et al. Estimated prevalence of anabolic steroid use among male high school seniors. *JAMA.* 1988;260:3441−3445.

15. Yesalis CE, Kennedy NJ, Kopstein AN, et al. Anabolic-androgenic steroid use in the United States. *JAMA.* 1993;270(10):1217−1221.

16. Albertson TE, Chenoweth JA, Colby DK, et al. The changing drug culture: use and misuse of appearance- and performance-enhancing drugs. *FP Essentials.* 2016;41:30−43.

17. Heinlein CA, Chang C. Androgen receptor (AR) coregulators: an overview. *Endocr Rev.* 2002;23:175−200.

18. Yanase T, Fan W. Modification of androgen receptor function by IGF-1 signaling implications in the mechanism of refractory prostate carcinoma. *Vitam Horm.* 2009;80:649−666.

19. Arnold AM, Peralta JM, Thonney ML. Ontogeny of growth hormone, insulin-like growth factor-I, estradiol and cortisol in the growing lamb: effect of testosterone. *J Endocrinol.* 1996;150:391−399.

20. Ting HJ, Yeh S, Nishimura K, et al. Supervillin associates with androgen receptor and modulates its transcriptional activity. *Proc Natl Acad Sci USA.* 2002;99:661−666.

21. Kraemer WJ. Endocrine response to resistance exercise. *Med Sci Sports Exerc.* 1988;20: S152−S157.

22. Ferrando AA, Stuart CA, Sheffield-Moore M, et al. Inactivity amplifies the catabolic response of skeletal muscle to cortisol. *J Clin Endocrinol Metab.* 1999;84:3515−3521.

23. Hackney AC. Stress and the neuroendocrine system: the role of exercise as a stressor and modifier of stress. *Expert Rev Endocrinol Metab.* 2006;1(6):783−792.

24. Hackney AC. *Exercise, Sport, and Bioanalytical Chemistry: Principles and Practice.* New York, New York: Elsevier — RTI Press; 2016.

25. Bhasin S, Storer TW, Berman N, et al. Testosterone replacement increases fat-free mass and muscle size in hypogonadal men. *J Clin Endocrinol Metab.* 1997;82:407−413.

26. Bhasin S, Woodhouse L, Casaburi R, et al. Testosterone dose-response relationship in healthy young men. *Am J Physiol Endocrinol Metab.* 2001;281:E1172−E1181.

27. Dolny DG, Hackney AC, Van Zanteen EL. Short-term effect of anabolic steroids and testosterone administration on serum lipoproteins, sex hormones and body composition. *Biol Sport.* 1992;9(1):25−32.

28. Shahani S, Braga-Basaria M, Maggio M, et al. Androgens and erythropoiesis: past and present. *J Endocrinol Invest.* 2009;32(8):704−716.

29. Kuipers H. Anabolic steroids: side effects. In: Fahey TD, ed. *Encyclopedia of Sports Medicine and Science.* Auckland: Internet Society for Sport Science: <http://www.sportsci.org/encyc/index.html>; 1998 (accessed 01.04.17).

30. Alen M, Rahkila P. Anabolic-androgenic steroids effects on endocrinology and lipid metabolism in athletes. *Sports Med.* 1988;6:327−332.

31. Cohen JC, Hickman R. Insulin resistance and diminished glucose tolerance in power lifters ingesting anabolic steroids. *J Clin Endocrinol Metab.* 1987;64:960−963.

32. DePiccoli B, Giada F, Benettin A, et al. Anabolic steroid use in body builders: an echocardiographic study of left ventricular morphology and function. *Int J Sports Med.* 1991;12:408−412.

33. Haupt HA. Anabolic steroids and growth hormone. *Am J Sports Med.* 1993;21:468−474.

34. Mauras N, Hayes V, Welch S, et al. Testosterone deficiency in young men: marked alterations in whole body protein kinetics, strength, and adiposity. *J Clin Endocrinol Metab.* 1998;83:1886−1892.

35. Wilson JD. The role of 5 alpha-reduction in steroid hormone physiology. *Reprod Fertil.* 2001;13:673−678.

36. El Osta R, Almont T, Diligent C, et al. Anabolic steroids abuse and male infertility. *Basic Clin Androl.* 2016;26:2.

37. Leder BZ, Longcope C, Catlin DH, et al. Oral androstenedione administration and serum testosterone concentrations in young men. *JAMA.* 2000;283:779−782.

38. Leder BZ, Catlin DH, Longcope C, et al. Metabolism of orally administered androstenedione in young men. *J Clin Endocrinol Metab.* 2001;86:3654−3658.

39. Brooks GA, Fahey TD, White TP. *Ergogenic aids. Exercise Physiology: Human Bioenergetics and Its Application.* Mountain View, CA: Mayfield Publishing; 1996:617−630.

40. Prough R, Clark BJ, Klinge CM. Novel mechanisms for DHEA action. *J Mol Endocrinol.* 2016;56(3):R139−R155.

41. Yen SS, Morales AJ, Khorram O. Replacement of DHEA in aging men and women: potential remedial effects. *Ann N Y Acad Sci.* 1995;774:128−142.

42. Armstrong J. East Germany's forgotten Olympic doping victims tell of illness, infertility and changing sex. *Mirror.* December 4, 2015. <http://www.mirror.co.uk/news/uk-news/east-germanys-forgotten-olympic-doping-6949436>; 2015 (accessed 01.07.17).

CHAPTER *3*

Stimulants

Stimulant drugs elevate mood, alertness, energy, stamina, and muscle power. These effects are mediated by influencing both the central nervous system and peripheral nervous systems. In a medical context, these effects can be therapeutic. In athletics, stimulants can enhance physical performance, consequently most of them are banned by the World Anti-Doping Agency (WADA) and other anti-doping authorities, though some are allowed under a therapeutic use exemption. The repeated use of stimulants can produce serious negative psychophysiological side effects such as paranoia, hostility, and addiction.

LOSING WEIGHT, STAYING AWAKE, GAINING FOCUS

Stimulants have a long history, having been used for religious, medicinal, and recreation purposes by many ancient societies and cultures.[1] These uses have persisted into modern times.[2] Pharmaceutically, stimulants were first extracted and synthesized in 1887 by Dr. Lazăr Edeleanu, a Romanian chemist working at the University of Berlin. He extracted a compound, ultimately to be known as amphetamine, from the Ma-Huang plant (*Ephedra sinica*) found in China, which contains ephedrine and pseudoephedrine.[1]

Starting in the the1920s, amphetamine and structurally related compounds were used to treat asthma and other respiratory problems, obesity, and neurological disorders. As their potential for abuse and addiction became apparent, medical use declined, but recreational use is a major problem, and off-label prescription continues. Amphetamines are still prescribed for attention-deficit hyperactivity disorder, narcolepsy, and occasionally depression in patients who have not responded to standard treatments.[3] Ephedrine is a common over-the-counter stimulant used for asthma (acts as a bronchodilator), narcolepsy, obesity, and nasal congestion.[1,3]

Doping, Performance-Enhancing Drugs, and Hormones in Sport. DOI: https://doi.org/10.1016/B978-0-12-813442-9.00003-1

Caffeine use is socially acceptable and ubiquitous as a stimulant. It is found in soft drinks, foods, and over-the-counter products for weight loss and alertness. It has been consumed for millennia in coffee, tea, and cocoa, which contains the naturally occurring methylxanthine from which caffeine, theophylline, and theobromine are derived.[2,4]

Table 3.1 shows the stimulants on the United States Drug Enforcement Agency (DEA) list of controlled substances, meaning that dispensing and possession are regulated by the federal government. A substance is placed into one of five schedules (i.e., categories) according to whether it has a currently accepted medical use, its relative abuse potential, and its likelihood of causing dependence.[5] Caffeine and related compounds are unscheduled and therefore not regulated as drugs by the DEA or the Food and Drug Administration.[3,5] A detailed account of the degree to which each scheduled stimulant is thought to be used by athletes is beyond the scope of this book, but some idea is given in the following discussion.

STAYING ATHLETICALLY LEAN AND ON TASK

Stimulants are powerful activators–facilitators of the central and peripheral nervous systems, that is, sympathomimetic agents. The sympathomimetic action induces physiological changes such as increasing heart rate, blood pressure, body temperature, and energy metabolism.[6] These changes positively affect psychophysiological aspects of preparing for and dealing with the physical stress of exercise training and sports competition.[7] In short, athletes use stimulants to reduce fatigue and increase alertness, focus, competitiveness, and aggressiveness, and to alter some aspects of metabolism.[8] The purpose of this last point, altering metabolism, is typically to lose weight and change aspects of body composition (see the close-up at the end of this chapter). As to aggressiveness, armed services began using amphetamine in World War II, and the practice appears to continue. For example, stores of stimulants have been frequently found among Islamic terrorists fighting for Al-Qaeda and affiliated groups.[9,10] Anecdotally, National Football League and Major League Baseball players report having liberal access to stimulants throughout the 1960s and 70s.[10]

The most common stimulants associated with sport are caffeine and amphetamines, which is why this chapter targets them.[8] Over-the-counter

Table 3.1 Stimulants Scheduled by the United States Drug Enforcement Agency (trade names in parentheses)[5]

Schedule I

No accepted medical use in the US and high abuse potential. Cannot be prescribed

- aminoxaphen (Aminorex)
- amphetamine variants [e.g., 3,4-methylenedioxymethamphetamine (MDMA, "ecstasy")]
- cathinones
- fenethylline
- methcathinone
- mephedrone
- methylaminorex

Schedule II

High abuse potential with severe psychic or physical dependence. Narcotics, stimulants, and depressants. Written prescription required, signed by the practitioner, no renewals

- cocaine
- dextroamphetamine (Dexedrine)
- lisdexamfetamine dimesylate (Vyvanse)
- methamphetamine (Desoxyn)
- methylphenidate (Ritalin)
- phenmetrazine (Preludin)
- biphetamine

Schedule III

Abuse potential less than Schedules I and II. Prescription oral or written, up to five renewals within 6 months

- benzphetamine (Didrex)
- chlorphentermine
- clortermine
- phendimetrazine tartrate (Plegine, Prelu 2)

Schedule IV

Abuse potential less than Schedule III. Prescription oral or written, up to five renewals within 6 months

- armodafinil (Nuvigil)
- diethylpropion hydrochloride (Tenuate)
- fencamfamin
- fenproporex
- mazindol (Sanorex, Mazanor)
- mefenorex
- modafinil (Provigil)
- norpseudoephedrine
- phentermine (Fastin, Ionamin, Adipex)
- pipradrol
- sibutramine (Meridia)

Schedule V

Abuse potential less than Schedule IV. Subject to state and local regulation, and prescription may not be required

- pyrovalerone

and behind-the-counter remedies for colds and allergies contain ephedrine, pseudoephedrine hydrochloride (Sudafed), and phenylpropanolamine, which are structurally related to amphetamine but have legitimate therapeutic use and are far less associated with sports doping. Athletes also use illicit street drugs such as cocaine and methamphetamine, but more for recreation than performance enhancement.[3,8]

Prescription stimulants and some of their derivatives found in over-the-counter medications are on the WADA list of banned substances. Caffeine is not now banned by WADA but has been in the past. In 2017, caffeine was added to WADA's monitoring program so that scientific experts could study whether athletes were using it "with the intent of enhancing performance." Hence, it might return to the list if deemed to be used extensively as a performance-enhancing drug.[11]

NEUROCHEMISTRY

Amphetamine and its analogs exert their short-term effects by modifying communication among brain neurons. They cause release of dopamine and norepinephrine from presynaptic vesicles into the synapse and block the transport of dopamine from the synaptic space back into the presynaptic neuron (reuptake). Excessive dopamine in the synapse overrides the chemical breakdown process of the neurotransmitter at the synapse, which prolongs the stimulation of postsynaptic receptors.[12] This means that the chemical-based communication signaling between the neurons is extended. The overt organismal results are mood elevation (elation, euphoria) and hyperkinesia (increased motor activity).[13] Animal and human studies, notably via brain imaging, suggest that the chronic use of amphetamine-like stimulants changes the structure and function of dopaminergic neurons in the limbic reward system (ventral tegmentum and nucleus accumbens regions of the brain), an effect proposed to underlie the development of addiction.[14] In animal studies, high doses of dopaminergic stimulants produced permanent neurotoxic effects by damaging neuron cell endings.[15] Whether similar direct neurotoxic effects occur in humans is an issue of some debate, but it seems probable.

Caffeine and its congeners, which produce a less powerful stimulation than amphetamines, act by a different neurochemical mechanism. Caffeine reversibly blocks the action of adenosine on adrenergic receptors in neurons and peripheral systemic tissues.[16] Fig. 3.1 shows the widespread impact of this antagonism. An important element relative to energy metabolism is that caffeine increases lipolysis, a hydrolysis of triglycerides into their component parts, glycerol and fatty acids.[6] This leads to elevated glycerol and free fatty acid levels in blood plasma and increased fat metabolism.[17] This effect is induced as the inhibition of adenosine receptors increases intracellular levels of the secondary messenger cascade mechanisms via 3′,5′-cyclic adenosine monophosphate, leading to activation of hormone-sensitive lipase, an initial promoter of the lipolysis triglyceride hydrolysis.[6] Enhanced lipid metabolism via lipolysis as an energy sources during some forms of

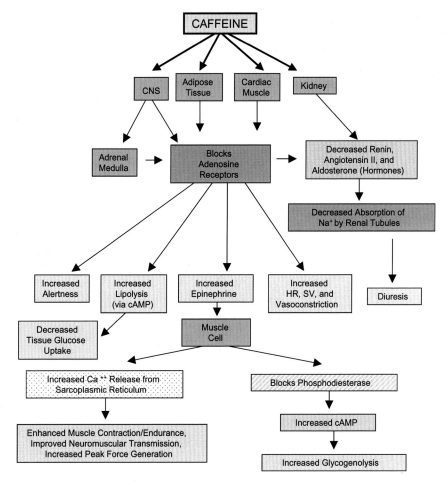

Figure 3.1 Major effects of caffeine on tissues and physiologic systems. CNS, central nervous system; HR, heart rate; SV, stroke volume (amount of blood pumped per heartbeat); cAMP, cyclic 3′, 5′ adenosine monophosphate (intracellular signaling agent).

exercise is advantageous, as it spares the limited carbohydrate stores (liver and muscle glycogen) of the body. This is crucial, because depletion of glycogen is a potential causative factor in muscular fatigue.[6]

PERFORMANCE ENHANCEMENT EVIDENCE

A 2017 survey by BBC Sport of over 1000 British adult athletes found that 14% had taken some form of strong stimulant such as amphetamine during their career.[18] These data agree with a survey of approximately 12,000 National Collegiate Athletic Association (NCAA) university athletes in the United States in 2013.[19] The percentage for caffeine usage is of course far higher because of its prevalence in food and drink. Mitchell et al.[20] found that 85% of the United States population (including high school and NCAA level athletes) consumed at least one caffeine beverage per day. These statistics are applicable worldwide.[2,3,4]

Amphetamine

Exercise studies have shown significant, dose-dependent ergogenic effects of amphetamine, although experiments are limited by the ethical constraints.[21] Work in 1940 by Heyrodt and Weissenstein showed that runs to exhaustion could be extended.[22] In 1947, Cuthbertson and Knox reported in a case study that pedaling time to exhaustion on a cycle ergometer could be extended for several hours.[23] Work by Chandler and Blair in 1980 supported these earlier findings.[24] Not all of the early studies found an ergogenic effect; these failures could be attributed to the dosing and testing protocol variations.[10,21]

Zaretsky et al.[25] found higher maximal oxygen uptake (VO_{2max}, aerobic energy capacity) and run time to exhaustion in rats, confirming an effect found in humans.[21,24] The improvement seems to come from masking fatigue, increasing alertness, and stimulating the sympathomimetic system—that is, producing physiological effects characteristic of sympathetic nervous system activation mediated by catecholamines. The activation of this system has profound effects on the cardiovascular system and energy metabolism, with increased heart rate and myocardial contractility as well as increased lipolysis leading to enhanced endurance capacity.[6] Amphetamines are on the WADA banned list, of course, but a therapeutic use exemption can be granted with proper medical documentation of medical etiology and accepted standard of care practices (see Chapter 1: Overview: Doping in Sport).[26]

As an aside, like amphetamine, cocaine can mask fatigue and pain and increase alertness. Because it is addictive, ethical constraints have prevented large-scale research on exercise ergogenic effects. Avois and associates conclude that, contrary to popular belief, it has no ergogenic value.[27] It can be detrimental when combined with exercise; there are case study reports of cardiovascular-related exercise deaths with its use.[28] It too is banned by WADA, as well as being an illegal recreational drug in most of the world.

Caffeine

As a component of tea, coffee, cocoa, soft drinks, some foods, and over-the-counter pharmaceuticals, caffeine is a part of daily life and one of the most studied ergogenic substances. Acutely, it improves aerobic exercise performance, including high-intensity short-duration bouts, and overall muscular strength and power, as well as some aspects of cognitive ability.[2,4,29] These benefits are clearly documented in many studies involving running, swimming, and cycling.[2,4] The effects come through influencing energy substrate actions—increased lipolysis and glycogen sparing ("metabolic effect")—and central nervous system stimulation of motor neurons and their muscle fibers ("nerve effect"). The metabolic effect appears highly beneficial in prolonged exercise, while the nerve effect enhances activities involving strength and power.[2,4] As with other stimulants, there are inconsistencies in research findings, with some studies reporting little or no benefit. These contradictions are attributed to dosing and testing protocol variances as well as tolerance development and sensitivity to caffeine.[10] Tolerance is acquired over time with repeated exposure and results in a need for higher doses to invoke responses, while sensitivity refers to one's genetic predisposition to metabolism of and responding to caffeine.[2,4,29]

The physiological responses are dose dependent, and extremely high doses can create a toxicity reaction. Also, there is wide variance between people in responsiveness, which could be related to tolerance in habitual consumers. There is some evidence that abstention for at least 7 days restores optimal benefit in acute use.[30] One negative effect is diuresis, which increases the risk of dehydration, which can compromise performance and increase the risk of heat injury.[6] Caffeine has a varied history of regulation by antidoping agencies. The International Olympic Committee initially banned it in the 1960s, lifted the ban in the 1970s, and then banned high levels of use in the 1980s. After much back and forth, it is currently not on the WADA Prohibited List but is on the monitoring program list of substances that WADA follows to detect patterns of misuse (see earlier comments).[11]

SPEEDING AND CRASHING

Stimulant intoxication produces characteristic signs and symptoms that depend on the potency of the agent (amphetamine > caffeine) and dose:[2,4]

- tension, anxiety, restlessness, agitation
- dyskinesia (abnormality or impairment of voluntary movement)
- euphoria, irritability, lability of mood
- talkativeness
- variable or random mental processes (confusion, tangential thoughts, paranoia, hallucinations, amnesia)
- impaired insight and judgment
- improved alertness (at small doses)

Hyperthermia, hyponatremia, and cardiac arrhythmias are side effects which collectively increase the risk for myocardial infarction and hemorrhagic stroke, with life-threatening consequences.

Stimulant withdrawal can produce symptoms that may compromise the ability to function:

- sedation
- hypoactivity
- depressed or irritable mood
- inattentiveness
- taciturnity (being reserved or reticent)
- paranoia
- impaired insight and judgment
- fatigue

CONCLUSION

The most prevalent stimulants used by athletes are amphetamine and caffeine. Both can enhance performance, but not in all exercise situations or all sports, and both have potentially dangerous psychic and physiologic side effects at high doses. Amphetamine's sympathomimetic action can produce hyperthermia, hyponatremia, and cardiac arrhythmias, which increase the risk for myocardial infarction and hemorrhagic stroke. Repeated use can produce physiological and psychological dependence. Amphetamines are banned by WADA but caffeine is not, although caffeine is currently under review and being considered for banning.

Close-Up: Weight Loss and Body Composition—What's the Difference Between Lean Body Mass, Fat-Free Mass, and Muscle Mass, and Why Do Athletes Care?

Kristin S. Ondrak
American Public University, Charles Town, WV, United States

Athletes frequently take stimulants to aid in weight loss and alter their body composition. But why is the makeup so important to them? A person's body weight (mass) is comprised of numerous components, commonly quantified via measures of body composition. These compositional results are used for classification of disease risk, tracking progress in fitness programs, and estimating performance in athletic events, among other things.

Three commonly used body composition terms that are confusing are lean body mass (LBM), fat-free mass (FFM), and muscle mass, and the goal of this close-up is to compare and contrast them. LBM is the most comprehensive of the three, as it includes all of the following: muscle tissues, bones, connective tissues, water, internal organs, as well as essential fat, which is in bone marrow and surrounds internal organs. Essential fat is necessary for health and should not be targeted for reduction in the way that is done with storage depot fat (adipose tissue). As depicted in Fig. 3.2, storage fat is the only component of one's body mass that is excluded from LBM. FFM, on the other hand, includes the same components as LBM with the exception of essential fat. Muscle mass is defined solely as skeletal muscle tissue; in many sporting activities, the extent of its development is critical. The percentage of one's mass attributed to these components depends on many factors such as age, sex, genetics, diet, and type and degree of exercise training.

Why are these components important to athletes? Body composition alone cannot predict success in athletics, but it certainly plays a key role

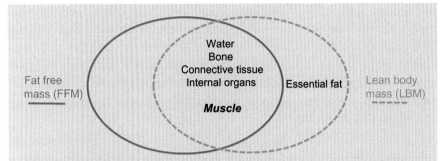

Figure 3.2 The components of fat-free mass (FFM) and lean body mass (LBM) in the human body. (N.B.: Storage fat is not included in FFM or LBM.)

in many sports. For example, some athletes strive to maximize their LBM and FFM and minimize storage fat, specifically in sports where competitors compete in weight classes (e.g., wrestling and martial arts) or in sports where physique and appearance may play a role in scoring (e.g., diving, figure skating, and gymnastics). In sports where strength or anaerobic (high intensity, short duration, "power") bursts are important, such as some football positions and weight lifting, athletes strive to maximize muscle mass and LBM. For example, researchers reported a strong and significant relationship between LBM and predicted sprint performance in a study of female cross-country skiers.[31] While LBM may be favorable for an athlete's anaerobic capacity, high amounts of LBM can be related to reduced aerobic capacity. In fact, researchers showed that relative VO_{2max} (mL/kg/min) was ~9% lower for men with a high amount of lean mass compared to men with a similar percent body fat but less lean mass.[32] This suggests that having too much mass, even if it's "favorable" mass, can be harmful to one's ability to excel in aerobic events. Finally, in sports with a heavy aerobic component such as endurance running, the volume and type of training often reduce total body mass and storage fat to very low levels. The amount of muscle mass and LBM will then depend on the resistance training component, if any, of their program. Taken together, the relative importance of LBM, FFM, and muscle mass are dependent on the volume and type of training undertaken and are ultimately governed by the demands of the sport.

REFERENCES

1. Crocq MA. Historical cultural aspects of man's relationship with addictive drugs. *Dialogues Clin Neurosci.* 2007;9(4):355–361.

2. Paluska SA. Caffeine and exercise. *Curr Sports Med Rep.* 2003;2:313–319.

3. Misuse of Prescription Drugs – What are Stimulants? NIH – NIDA. <https://www.drugabuse.gov/publications/research-reports/prescription-drugs/stimulants/what-are-stimulants>; 2016 (accessed 03.12.2017).

4. Keisler BD, Arnsey TD. Caffeine as an ergogenic aid. *Curr Sports Med Rep.* 2006;5:215–219.

5. United States DEA Diversion Control. *Controlled Substance Schedules.* DEA Diversion Control. Available at <http://www.deadiversion.usdoj.gov/schedules/orangebook/orangebook.pdf>; 2016 (accessed 04.02.2017).

6. Hackney AC. *Exercise, Sport and Bioanalytical Chemistry: Principles and Practice.* New York, New York: Elsevier-RTI Press; 2016.

7. Hackney AC. Stress and the neuroendocrine system: the role of exercise as a stressor and modifier of stress. *Expert Rev Endocrinol Metab.* 2006;1(6):783–792.

8. Australian Sport Anti-Doping Authority. Important Athlete Advisory: Prohibited Stimulants Found in Supplements. <www.asada.gov.au>; 2014 (accessed 02.28.2017).

9. Henley J. Captagon: the amphetamine fueling Syria's civil war. *Guardian Newspaper.* January 13, 2014. <https://www.theguardian.com/world/shortcuts/2014/jan/13/captagon-amphetamine-syria-war-middle-east>; 2014 (accessed 07.20.2017).

10. Strauss R. *Drugs and Performance in Sport.* New York: WB Saunders Co.; 1987.

11. Bowden A. Could Caffeine be Headed Back Onto WADA's Banned List? March 10 2017. <http://road.cc/content/news/218871-could-caffeine-be-headed-back-wadas-banned-list>; 2017 (accessed 07.20.17).

12. Cooper J, Bloom F, Roth R. *The Biochemical Basis of Neuropharmacology.* 6th ed New York: Oxford University Press; 1991.

13. Treatment Improvement Protocol (TIP) Series, No. 33. Treatment for Stimulant Use Disorders. Center for Substance Abuse Treatment. Rockville, MD: Substance Abuse and Mental Health Services Administration—US; Publication ID: SMA09-4209; June 2009.

14. Self D, Nestler E. Molecular mechanisms of drug reinforcement and addiction. *Annu Rev Neurosci.* 1995;18:463–495.

15. Selden LS. Neurotoxicity of methamphetamine: mechanisms of action and issues related to aging. In: Miller MA, Kozel NJ, eds. *Methamphetamine Abuse: Epidemiologic Issues and Implications.* NIDA Research Monograph Series, Number 115. DHHS Pub. No. (ADM) 91-1836. Rockville, MD: National Institute on Drug Abuse; 1991: 24–32.

16. Nehlig A, Daval JL, Debry G. Caffeine and the central nervous system: mechanisms of action, biochemical, metabolic and psychostimulant effects. *Brain Res.* 1992;17(2):139–170.

17. Costill DL, Dalsky GP, Fink WJ. Effects of caffeine ingestion on metabolism and exercise performance. *Med Sci Sports.* 1978;10(3):155–158.

18. Sapstead N. BBC—Doping in Sport: Drug Use 'Fast Becoming a Crisis'. March 20, 2017. <http://www.bbc.com/sport/38884801>; 2017 (accessed 07.01.2017).

19. Buckman JF, Farris SG, Yusko DA. A national study of substance use behaviors among NCAA male athletes who use banned performance enhancing substances. *Drug Alcohol Depend.* 2013;131(0):50–55.

20. Mitchell D, Knight C, Hockenberry J, et al. Beverage caffeine intakes in the U.S. *Food Chem Toxicol.* 2014;63:136–142.

21. Ivy J. Amphetamines. In: Williams MH, ed. *Ergogenic Aids in Sport.* Champaign, IL: Human Kinetics; 1983.

22. Heyrodt H, Weissenstein J. Uber steigerung korperlicher leistungfahigheit durch pervitin. *Arch Exp Pathol Pharmakol.* 1940;195:273–275.

23. Cuthbertson DP, Knox JAC. The effects of analeptics on the fatigued subject. *J Physiol (London)*. 1947;106:42–58.

24. Chandler JB, Blair VS. The effects of amphetamines on the select physiological components related to athletic success. *Med Sci Sports Exerc.* 1980;12(1):65–69.

25. Zaretsky DV, Brown MB, Zaretskaia MV, et al. The ergogenic effect of amphetamine. *Temperature (Austin)*. 2014;1(3):242–247.

26. WADA—2017 List of Prohibited Substances and Methods. <https://www.wada-ama.org/en/prohibited-list>; 2017 (accessed 07.20.17).

27. Avois L, Robinson N, Saudan C, et al. Central nervous system stimulants and sport practice. *Br J Sports Med.* 2006;40(Suppl 1):i16–i20.

28. Cantwell JD, Rose FD. Cocaine and cardiovascular events. *Physician Sport Med.* 1981;14(11):77–82.

29. Powers SK, Dodd SL. Caffeine and endurance performance. *Sports Med.* 1985;2:165–174.

30. Dodd SL, Brooks E, Powers SK, et al. The effect of caffeine on graded exercise performance in caffeine naïve versus habituated subjects. *Eur J Appl Physiol.* 1991;62:424–429.

31. Carlsson T, Tonkonogi M, Carlsson M. Aerobic power and lean mass are indicators of competitive sprint performance among elite female cross-country skiers. *Open Access J Sports Med.* 2016;7:153–160.

32. Maciejczyk M, Wiecek M, Szymura J, et al. The influence of increased body fat or lean body mass on aerobic performance. *PLoS ONE.* 2014;9(4):e95797.

CHAPTER 4

Glucocorticoids

Glucocorticoids belong to the corticosteroid classification of chemical compounds and in humans are hormones released by the adrenal cortex. They are essential for normal life, as they regulate or support physiological processes in the cardiovascular, metabolic, immunologic, and other systems. In a medical context, they are powerful antiinflammatory agents, which results in their widespread clinical use. Of the endogenous glucocorticoids, perhaps the most critical is cortisol, because it affects so many of the body's physiological systems. The wide ranging and powerful effects of glucocorticoids make them readily abusable as performance-enhancing drugs (PEDs) in the sporting world.

ANTIINFLAMMATORY AND PULMONARY ACTIONS

The English physician Thomas Addison recognized the importance of glucocorticoids in physiological function and published his results as *On the Constitutional and Local Effects of Disease of the Suprarenal Capsules* (1855). This short monograph contains the classic description of the endocrine disturbance now known as Addison's disease, in which the adrenal glands (capsules) do not produce enough glucocorticoids. He established that the adrenal glands are essential for a healthy life. Nearly a century later, adrenal glucocorticoid secretions were isolated and identified by the American chemist Edward Kendall of the Mayo Clinic, for which, he was awarded the 1950 Nobel Prize in Physiology or Medicine, along with Philip Hench and Tadeusz Reichstein, who made similar discoveries.[1,2]

There is a disease that is the exact opposite of Addison's, what is now called Cushing's syndrome, and it is caused by glucocorticoid excess and was studied by many of Addison's contemporaries. It was not until 1932, however, that American neurosurgeon Harvey Cushing described the clinical findings that provided the link between physical signs (e.g., abnormal obesity of the face and trunk) and a specific type of pituitary tumor that led to excess glucocorticoid secretion. Later, it

Doping, Performance-Enhancing Drugs, and Hormones in Sport. DOI: https://doi.org/10.1016/B978-0-12-813442-9.00004-3

became clear that many patients with similar symptoms and signs did not have this tumor. So the use of the term Cushing's syndrome was modified to refer to all patients with the classic symptoms and signs, regardless of the cause, while the term Cushing's disease is restricted to patients in whom the cause is excess adrenocorticotropic hormone (ACTH) secreted by an anterior pituitary tumor.[1,2] The extensive historic medical research on the pathologies of Addison's and Cushing's disease led to much of our understanding of how glucocorticoids work.

In humans, a variety of glucocorticoids have been identified as products of the adrenal cortex glands. Cortisol is the most plentiful and more biologically active one, while 11-deoxy-cortisol, 17-OH-progesterone, and 17-OH-pregnenolone are less widely secreted and of lower activity, though important.[3,4]

Clinically, glucocorticoids are administered topically and orally for several medical conditions. Cortisol, called hydrocortisone when used as medicine, is applied topically for insect bites, poison oak, poison ivy, eczema, dermatitis, allergies, and unspecified rash. It mitigates the local inflammation response such as swelling, itching, and redness. Oral glucocorticoids are used for certain disorders of the blood, endocrine system, immune system, skin and eye irritations, arthritis, breathing problems, aspects of cancer, and severe allergies. Oral administration reduces the entire immune system's response and must be used with caution. It is also appropriate for diseases of the adrenal gland such as Addison's, where there is adrenocortical insufficiency resulting in reduced glucocorticoid levels in the blood.[1-3] Injectable of this drug have similar effects, but more localized actions. Table 4.1 shows some commonly prescribed glucocorticoids.[4]

Table 4.1 Commonly Prescribed Glucocorticoid Drugs[4]
• **Prednisone and Prednisolone**—Most commonly used examples because of high glucocorticoid activity. Prednisone is transformed by the liver into prednisolone. Prednisolone may be administered in tablet form or produced by the body from prednisone. These medications are often considered interchangeable in medical practice. Common trade names are Deltasone, Rayos, Prednicot, and Sterapred.
• **Dexamethasone**—Has a particularly high glucocorticoid activity and low mineralocorticoid[a] activity and so can be used in higher doses. It is often prescribed to reduce nerve swelling after neurological trauma and neurosurgery. Common trade names are Ozurdex, Baycadron, and Maxidex.
• **Hydrocortisone**—Has much more mineralocorticoid activity than prednisone and is therefore usually not suitable for long-term use internally. Externally, it is used extensively as a cream or lotion for skin conditions such as rash and itching. Common trade names are Hydrocortone, Cortef, Solu-Cortef, and Colocort.
[a]*Mineralocorticoids are corticosteroids. An example is the hormone aldosterone, which is involved with maintaining electrolyte and water balance.*

Glucocorticoids can also be inhaled and used to suppress airway inflammation. This antiinflammatory action occurs by activating anti-inflammatory genes, switching off inflammatory gene expression, and inhibiting inflammatory cell response (see later discussion).[5] In addition, they enhance beta-2 adrenergic signaling by increasing beta-2 receptor expression and function (see Chapter 6: Beta-2 Agonists). The effect of these actions is to lessen the symptoms and signs of asthma in most patients and hence allow them to breathe more freely.[3]

HELPING ATHLETES RECOVER FASTER, BREATHE BETTER, AND BURN MORE FAT

Glucocorticoids are widely used in the management of sports-related injuries as well as disorders caused by overuse of muscles and muscle–tendon junctions—that is, for antiinflammatory effect (see the close-up at the end of this chapter). These problems occur often in athletes due to the demands of strenuous training regimens. Some informal surveys suggest that glucocorticoids could be the most prevalent drugs used by athletes,[6] in part because they are common in medical practice and relatively easy to obtain.[3,4] For example, cortisone is a bronchodilator prescribed for asthmatic athletes to aid breathing (see Chapter 6 for more detail). Diagnosis and prescription require confirmation by an independent physician before an athlete can be considered for a therapeutic use exemption (see Chapter 1: Overview: Doping in Sport). However, evidence clearly indicates that glucocorticoids are also used as ergogenic PEDs because they reduce fatigue centrally and peripherally and increase energy substrate mobilization and metabolism.[7,8]

Oral, intravenous, intramuscular, and rectal administration of glucocorticoids in competition is prohibited by the World Anti-Doping Agency. All other modes of application are permitted without restriction if for therapeutic purposes. Out-of-competition use also is permitted, so no application for a therapeutic use exemption needs to be submitted.

ENDOCRINE AND IMMUNE ACTIONS

Hormonally cortisol has widespread physiological effects (Fig. 4.1). Its production is controlled by a series of endocrine glands called the

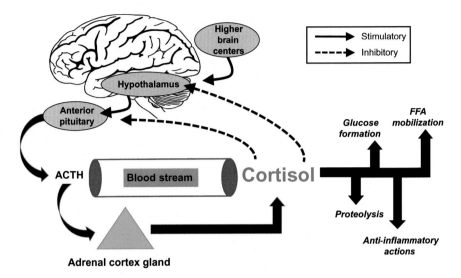

Figure 4.1 The pathways of cortisol's regulation and key physiological effects in the body. FFA, *free fatty acids, denoting mobilization and increased lipolysis.*

hypothalamic–pituitary–adrenal (HPA) axis. Signaling stimuli to the periventricular nucleus area of the hypothalamus, such as hypoglycemia or physical or emotional stress, results in the secretion of corticotropin-releasing hormone into the hypophyseal portal vascular system. Corticotropin-releasing hormone then stimulates the release of ACTH from the anterior pituitary gland. ACTH in turn causes the synthesis and secretion of cortisol by the adrenal cortex.

Approximately 90% of the secreted cortisol is bound to corticosteroid-binding globulins, which are carrier proteins in the blood. Free unbound cortisol circulating in the blood is the biologically active form of the hormone which can bind to glucocorticoid receptors (GRs) and initiate physiological events. GRs are expressed in virtually all human cells. At many tissues, cortisol is converted to its less active form, cortisone, by the enzyme11β-hydroxysteroid dehydrogenase. This enzymatic conversion is bidirectional and is a cellular method of regulating GR activation. The GR is a member of the steroid hormone receptor family of proteins which bind cortisol with high affinity; the receptor-bound cortisol promotes the dissociation of chemical structures called chaperones, including heat-shock proteins, from the receptor in the activation process.[5,9]

Within the cell, cortisol acts in three ways to influence physiological function of tissues:

- The cortisol-GR complex moves to the nucleus, where it binds to a segment of cellular DNA called the glucocorticoid-responsive elements. The resulting complex recruits proteins (coactivator or corepressor) that modify the structure of chromatin, thereby facilitating or inhibiting assembly of the basal transcription process and initiation of transcription by RNA. This is the classic genomic process by which steroid hormones function (i.e., direct gene activation mechanism).
- Regulation of other glucocorticoid-responsive genes involves interaction between the cortisol-GR complex and other transcription factors such as nuclear factor-κB (NF-κB).[5] This mechanism seems to act at lower cortisol levels than the events noted above to change transcription. This is evidence that the mechanism of action is a dose-dependent reaction.
- Glucocorticoid signaling through membrane-associated receptors and secondary messengers (nongenomic pathways) also exists and is an alternative means to initiate physiological effects.[3,5,9,10]

The GR inhibits inflammation through all three of those mechanisms: direct genomic, indirect genomic, and nongenomic. Furthermore, interactions among the nervous system, the HPA axis, and components of the innate and adaptive immune system play a key role in the regulation of inflammation and immunity. For instance, cytokines and inflammatory mediators activate peripheral pain receptors, which project pain signals to the thalamus and the somatosensory cortex of the brain, leading to HPA activity. Glucocorticoids inhibit the synthesis of cytokines and inflammatory mediators, thus forming a negative feedback loop to lessen the response (Fig. 4.2). Select cytokines can also act directly on the brain to activate the HPA axis. Dysregulation of this cytokine−neuroendocrine loop by hyperactivity or hypoactivity of the HPA axis leads to changes in circulating cortisol levels, causing systemic changes in inflammation and immunity.[3,5,9,10]

- Hyperactivity of the axis in the absence of inflammation (e.g., Cushing's syndrome) causes immunosuppression and increased susceptibility to infection. Physical pain, emotional trauma, and caloric restriction also acutely activate the axis and can cause immunosuppression.

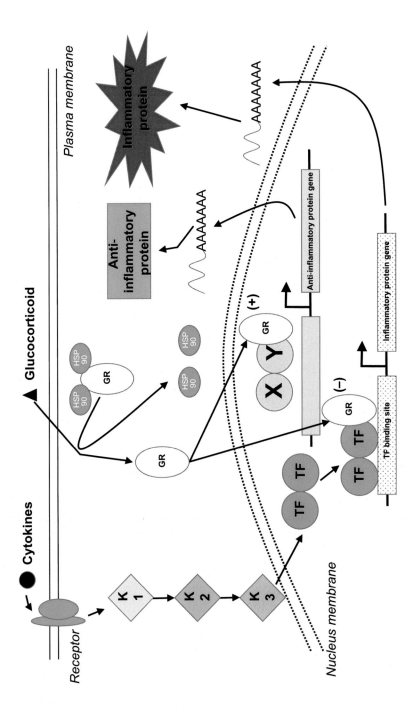

Figure 4.2 A generalized model of the inflammatory cascade involving glucocorticoids. Cytokine binding to receptors in the plasma membrane leads to activation of a kinase cascade (K1, K2, and K3). K3 translocates across the nuclear membrane and activates the transcription factor (TF) which induces transcription of an inflammatory protein gene leading to synthesis of the inflammatory proteins. Binding of glucocorticoid to the glucocorticoid receptor (GR) leads to dissociation of the heat-shock proteins (HSP90) and translocation of GR to the nucleus. Subsequently, GR may bind TF (e.g., NF-κB or AP-1) to repress activated transcription or interact with other factors (X, Y) to activate gene transcription of antiinflammatory genes promoting repressed translation of inflammatory genes.[9,10]

- Hypoactivity of the axis increases the susceptibility to and severity of many inflammation states, both severe and mild. In some such conditions, supplemental glucocorticoids may be necessary during an infection or inflammation to prevent the harmful effects of some cytokines produced in the body.

Interestingly, glucocorticoid resistance is common in patients with severe rheumatoid arthritis, usually due to decreased expression of GRs or activation of cellular mechanisms, which inhibits glucocorticoid signaling within cells.[3,5,9,10]

DO THEY WORK? YES, NO, AND MAYBE

Although glucocorticoids appear to be widely abused by athletes, there are fewer published studies looking at its ergogenic effect than with other PEDs. Studies by Petrides et al. and Marquet et al. found no greater effect than placebo on maximal oxygen uptake, a key index of aerobic capacity.[8,11] Nor was any influence found on prolonged endurance exercise to exhaustion in a fixed distance time trial or a series of brief high-effort exercises that mimic a competitive sporting situation.[12-14] Researchers had hypothesized that glucocorticoids might attenuate the fatigue and pain associated with exercise and thus enhance performance. However, these studies were limited in that, for ethical reasons, the doses were insufficient to push blood cortisol levels beyond the physiological range, while most doping athletes try to achieve supraphysiological levels.[6] One study that used a relatively high dose for 1 week found large significant improvements in performance; the physiological mechanism was unclear, but the authors hypothesized alterations in brain serotonergic and dopaminergic activity.[15] Elevated serotonin levels are associated with increased fatigue,[16] and the high glucocorticoid dose may have suppressed brain serotonin. Soetens et al.[7] found that synthetic ACTH, which enhances endogenous glucocorticoid levels, increased free fatty acid (FFA) mobilization during exercise; that is a glycogen sparing phenomenon and can lessen the rate of muscular fatigue development.[16] LePanse et al. conducted one of the few studies to examine women athletes and glucocorticoids.[17] One week of daily administration of glucocorticoids resulted in prolonged exercise time (66.4 ± 8.4 vs 47.9 ± 6.7 min) at $\sim 75\%$ VO_{2max}. The mechanism was unclear, but the authors speculated that altered hormonal profiles facilitated a more lipid-based (i.e., substrate mobilization) exercise energy production, which is vital in endurance-based performance.[16]

Animal research, where high doses can be administered, has shown clear ergogenic effects, supporting a PED role. Gorostiaga et al. found that a single injection of cortisol acetate in rats 21 hours prior to treadmill testing increased glycogen content in liver and muscles, plasma FFAs, and running time to exhaustion, compared to placebo.[18] Capaccio et al. found that after 14 consecutive daily injections of cortisol acetate in rats at doses high enough to produce skeletal wasting, and despite muscle atrophy, the cortisol groups showed enhanced exercise performance with increased total run times during ~42 minutes of testing.[19]

Some researchers and athletes advocate the use of pharmaceutical glucocorticoids during the recovery days from competition events because of the well-attested antiinflammatory and metabolic effects, which reduce localize pain and enhance glycogen resynthesis and storage. These effects are clearly therapeutic and aid the athlete in preparation for the next intense training session or competition and certainly provide for an enhanced training stimulus.[6]

PSEUDO-CUSHING'S SYNDROME AND MINERALOCORTICOID ACTIONS

Oral glucocorticoids can have mild-to-moderate side effects such as nausea, heartburn, headache, dizziness, menstrual period changes, trouble sleeping, increased sweating, and acne. They also reduce immunity to infections and may raise blood glucose to the point of mild hyperglycemia. The high doses used as PEDs can produce more serious effects such as tiredness, swelling of ankles and feet, weight gain, vision problems, easy bruising and bleeding, puffy face, unusual hair growth, depression, mood swings, agitation, muscle pain and weakness, thinning skin, slow wound healing, and bone pain. The swelling and weight gain are associated with the mineralocorticoid actions of glucocorticoids such as cortisol. Mineralocorticoids are another class of corticosteroid hormones found in the body, with aldosterone being the principle one.[16] They activate an alternative set of mineralocorticoid receptors (MRs) which produce downstream actions of electrolyte and water retention.[16] The highly similar chemical structures of cortisol and aldosterone result in a slight degree of cross-reactivity and activation on each other's receptors, cortisol binding to MRs and aldosterone to GRs.

Side effects from topical use are less severe and include stinging, burning, irritation, dryness, and redness at the application site. Acne, unusual hair growth, "hair bumps" (folliculitis), skin thinning or discoloration, and stretch marks may also occur, though rarely. But severe effects can occur: bleeding at the site, swelling (especially of the face, tongue, and throat), dizziness, and trouble breathing.

As in Cushing's syndrome, muscle wasting and weakness can occur with prolonged use of glucocorticoids. These effects are muscle type specific, with Type IIb skeletal fibers (fast twitch) being more susceptible than Type I (slow twitch).[16] Some studies have reported that exercise training reduced or suppressed these side effects. There is evidence of glucocorticoid-induced inhibition of nitric oxide-dependent endothelial relaxation of blood vessels, leading to diminished blood flow and oxygen delivery. Interestingly, reduced nitric oxide availability over time can impair endothelial function, leading to hypertension and atherosclerosis, both of which are major cardiovascular complications associated with excessive glucocorticoid use. Supraphysiologic doses also induce production of free radicals by interfering with the mitochondrial electron transfer system. This interference impairs mitochondrial function and adenosine triphosphate (ATP) energy production.[6] Collectively, these side effects have strong potential to reduce performance in an exercise session and diminish the adaptive responses to exercise training.

CONCLUSION

The major glucocorticoid in humans, cortisol, is critical to physiological function and overall health. The PED ergogenic effect of glucocorticoids is supported by research, but only in a few studies, as many findings are ambiguous. That said, these drugs are nonetheless popular pharmaceutical PEDs because of their ability to mitigate the pain and discomfort associated with postexercise inflammation. This role allows athletes to maintain high levels of training and in turn promote enhanced physiological adaptation and, in specific situations, perhaps perform better in competition.

Close-Up: *Primum non nocere*—Glucocorticoids and an Ethical Dilemma in Sports Medicine

Barnett S. Frank
University of North Carolina, Chapel Hill, NC, United States

As medical professionals treating sports injury, we are constantly confronted with a clinical paradox. We are in pursuit of providing our athletes with the highest level of care while trying to return them to a level of high performance in the most efficient and effective manner. However, efficient and long-term patient benefit may not always coincide. As health practitioners, we are bound to *Primum non nocere*—"first, do no harm."[20] As clinicians, we are confronted with a conundrum. Is the possibility of inducing long-term harm while effectively reducing acute pain and improving function appropriate? Along these lines, glucocorticoid injections into musculoskeletal tissues represent one of sports medicine's most perplexing interventions. It was 1948 when the first dose of cortisone was injected into a rheumatoid arthritis patient. The treatment effects were of historic proportion, so much so that the researchers involved were awarded the 1950 Nobel prize in medicine for their "discoveries relating to the hormones of the adrenal cortex, their structure, and biological effects" for the effective treatment of pain and reduction of inflammation in vivo.[1] However, at the time of discovery, the long-term effects of cortisone on tissues were not known.

Today, we have a significant body of evidence suggesting that we may indeed be inducing irreversible long-term harm to a patient's tissue when we inject a glucocorticoid into a joint, a tendon sheath, or a skeletal muscle.[21–23] Do the cultural demands of the sports medicine field warrant risking future damage, or do the immediate benefits of pain reduction and increased function to permit sport and physical activity participation outweigh these risks? Furthermore, individuals frequently identify pain and reduced functionality as one of the primary barriers to their participating in health-enhancing physical activity.[24] Again, we are presented with a paradox: if we restrict an acutely effective treatment that would enable increased potential for physical activity participation, are we effectively contributing in some part to a population's inactivity levels, the number one contributor to all-cause mortality?[25]

In this context, the phrase "first, do no harm" needs to be considered. Is the operative portion of this medical ethics statement "FIRST," or is it the latter, "do no harm"? Many a patient or athlete in pain and clinician would agree that there is likely "first" no harm done (and indeed perceived benefit) shortly following a cortisone injection. However, as clinicians are we not also responsible for the long-term well-being of our patients? In seeking answers to these questions, we should truly consider glucocorticoid use in the sports medicine field as a classic example of the need for applied research and its translation into clinical practice. This seems acutely important, as recent consensus has established that there is unquestionably a high risk of committing long-term damage to musculo-skeletal tissue, for reported abuse in athletics by clinicians is high.[21–23] So, the question remains, as clinicians should we be concerned with being "first" or "doing no harm"?

REFERENCES

1. Glyn J. The discovery and early use of cortisone. *J R Soc Med.* 1998;91(10):513–517.

2. Rubin RP. A brief history of great discoveries in pharmacology: a celebration of the centennial anniversary of the founding of the American Society of Pharmacology and Experimental Therapeutics. *Pharmacol Rev.* 2007;59(4):289–359.

3. Van der Velden VH. Glucocorticoids: mechanisms of action and anti-inflammatory potential in asthma. *Mediat Inflamm.* 1998;7(4):229–237.

4. Glucocorticod Pharmacology, Tulane University, School of Medicine. <http://tmedweb.tulane.edu/pharmwiki/doku.php/glucocorticoid_pharmacology>; 2017 (accessed 03.03.2017).

5. Oakley RH, Cidlowski JA. The biology of the glucocorticoid receptor: new signaling mechanisms in health and disease. *J Allergy Clin Immunol.* 2013;132(5):1033–1044.

6. Duclos M. Evidence on ergogenic action of glucocorticoids as a doping agent risk. *Physician Sports Med.* 2010;383:121–127.

7. Soetens E, De MK, Hueting JE. No influence of ACTH on maximal performance. *Psychopharmacology.* 1995;118(3):260–266.

8. Marquet P, Lac G, Chassain AP, et al. Dexamethasone in resting and exercising men: effects on bioenergetics, minerals, and related hormones. *J Appl Physiol.* 1999;87(1):175–182.

9. Rhen T, Cidlowski JA. Anti-inflammatory action of glucocorticoids—new mechanisms for old drugs. *N Engl J Med.* 2005;353:1711–1723.

10. Newton R. Molecular mechanisms of glucocorticoid action: what is important? *Thorax.* 2000;55(7):603–613.

11. Petrides J, Gold PW, Mueller GP, et al. Marked differences in functioning of the hypothalamic–pituitary–adrenal axis between groups of men. *J Appl Physiol.* 1997;82(6):1979–1988.

12. Arlettaz A, Collomp K, Portier H, et al. Effects of acute prednisolone intake during intense submaximal exercise. *Int J Sports Med.* 2006;27(9):673–679.

13. Arlettaz A, Collomp K, Portier H, et al. Effects of acute prednisolone administration on exercise endurance and metabolism. *Br J Sports Med.* 2008;42(4):250–254.

14. Baume N, Steel G, Edwards T, et al. No variation of physical performance and perceived exertion after adrenal gland stimulation by synthetic ACTH (Synacthen) in cyclists. *Eur J Appl Physiol.* 2008;104(4):589–600.

15. Collomp K, Arlettaz A, Portier H, et al. Short-term glucocorticoid intake combined with intense training on performance and hormonal responses. *Br J Sports Med.* 2008;42 (12):983−988.

16. Hackney AC. *Exercise, Sport, and Bioanalytical Chemistry: Principles and Practice.* New York, New York: Elsevier−RTI Press; 2016.

17. Le Panse B, Thomasson R, Jollin L, et al. Short-term glucocorticoid intake improves exercise endurance in healthy recreationally trained women. *Eur J Appl Physiol.* 2009;107 (4):437−443.

18. Gorostiaga EM, Czerwinski SM, Hickson RC. Acute glucocorticoid effects on glycogen utilization, O_2 uptake, and endurance. *J Appl Physiol.* 1988;64(3):1098−1106.

19. Capaccio JA, Galassi TM, Hickson RC. Unaltered aerobic power and endurance following glucocorticoid-induced muscle atrophy. *Med Sci Sports Exerc.* 1985;17(3):380−384.

20. Sokol DK. "First do no harm" revisited. *BMJ.* 2013;347:f6426.

21. Abate M, Guelfi M, Pantalone A, et al. Therapeutic use of hormones on tendinopathies: a narrative review. *Muscles Ligaments Tendons J.* 2016;6(4):445−452.

22. Abate M, Salini V, Schiavone C, Andia I. Clinical benefits and drawbacks of local corticosteroids injections in tendinopathies. *Expert Opin Drug Saf.* 2017;16(3):341−349.

23. Dean BJF, Lostis E, Oakley T, Rombach I, Morrey ME, Carr AJ. The risks and benefits of glucocorticoid treatment for tendinopathy: a systematic review of the effects of local glucocorticoid on tendon. *Semin Arthritis Rheum.* 2014;43(4):570−576.

24. McPhail S, Schippers M. An evolving perspective on physical activity counselling by medical professionals. *BMC Fam Pract.* 2012;13(1):31.

25. Lee IM, Shiroma EJ, Lobelo F, et al. Effect of physical inactivity on major non-communicable diseases worldwide: an analysis of burden of disease and life expectancy. *Lancet (London, England)* 2012;380(9838):219−229.

Peptide—Protein Hormones

Endogenous endocrine hormones regulate a multitude of physiological processes, such as metabolism, growth and development, water balance, cardiovascular function, and reproduction.[1] Hormones are classified chemically into two groups, steroid—lipid soluble and peptide—protein based. This chapter is about the second category and their use as performance-enhancing drugs (PEDs). Because peptide—protein hormones and their related PEDs are so numerous (Table 5.1) and space in this volume is limited, only two key ones are discussed here, growth hormone and erythropoietin (EPO).

OFFSETTING ENDOCRINE DEFECTS AND DYSFUNCTIONS

Growth hormone and EPO are critical in regulating many aspects of physiology and therefore have important medical functions in many disease states. This led to intense research over the last century to understand their mechanisms of action and develop similarly acting pharmaceuticals.

Growth Hormone

Growth hormone is produced in and released from the anterior pituitary gland. This is why the first purified extracts, in the 1940s, were from human cadaver pituitary glands. In the 1950s, these extracts were shown to promote growth in animals and children with hypopituitarism. In 1962, researchers reported increased vigor and sense of well-being in women with hypopituitarism who received the extracts.[2] Cadavers continued to be the only source until the mid-1980s, when the first recombinant technology variety (methionyl human growth hormone) was developed. In recombinant procedures, a section of a DNA sequence from a different species is placed into a host organism, which then synthesizes the desired product to be harvested. This new source was timely, as cadaver-derived growth hormone was found to transmit Creutzfeldt—Jacob disease and was withdrawn from the

Doping, Performance-Enhancing Drugs, and Hormones in Sport. DOI: https://doi.org/10.1016/B978-0-12-813442-9.00005-5

Table 5.1 Peptide Hormones, Growth Factors, and Related Substances Prohibited by the World Anti-Doping Agency (WADA; Category S2)

1. Erythropoietin receptor agonists
1.1 Erythropoiesis-stimulating agents (ESAs)
darbepoietin (dEPO)
erythropoietin (EPO)
EPO-Fc
EPO-mimetic peptides (EMPs)   e.g., CNTO 530 and peginesatide
GATA inhibitors   e.g., K-11706
methoxy polyethylene glycol-epoetin beta (CERA)
transforming growth factor-β (TGF-β) inhibitors   e.g., sotatercept, luspatercept
1.2 Nonerythropoietic EPO receptor agonists[a]
ARA-290
asialo EPO
carbamylated EPO
2. Hypoxia-inducible factor (HIF) stabilizers[b]
cobalt
molidustat
roxadustat
HIF activators   e.g., argon and xenon
3. Chorionic gonadotrophin (CG) and luteinizing hormone (LH) and their releasing factors
buserelin
gonadorelin
leuprorelin (in males)
4. Corticotrophins and their releasing factors
corticorelin
5. Growth hormone and its releasing factors
growth hormone releasing hormone (GHRH), and its analogs   e.g., CJC-1295, sermorelin, and tesamorelin
growth hormone secretagogues (GHS)   e.g., ghrelin and ghrelin mimetics such as anamorelin and ipamorelin
growth hormone releasing peptides (GHRPs)   e.g., alexamorelin, GHRP-6, hexarelin, and pralmorelin (GHRP-2)
Other prohibited growth factors
fibroblast growth factors (FGFs)
hepatocyte growth factor (HGF)

(Continued)

Table 5.1 (Continued)
insulin-like growth factor 1 (IGF-1) and its analogs
mechano growth factors (MGFs)
platelet-derived growth factor (PDGF)
vascular-endothelial growth factor (VEGF)
Any other growth factor affecting muscle, tendon, or ligament protein synthesis/degradation, vascularization, energy utilization, regenerative capacity, or fiber type switching

[a]*An agonist is a chemical that binds to a receptor and activates the receptor to produce a biological response.*
[b]*These substances serve as erythropoiesis-stimulating agents (ESAs) originally developed as novel antianemia therapies.*

legitimate drugs market in 1985, although it appears to be still available on the black market in some countries.[2,3]

Growth hormone is medically used in children with growth hormone deficiency or insufficiency and poor growth due to renal failure, Turner syndrome (girls with a missing or defective X chromosome), Prader—Willi syndrome (usually due to uniparental disomy in chromosome 15), and those born extremely small for their gestational age with poor overall growth or stature development. In adults, it is approved for AIDS-related tissue wasting and growth hormone deficiency, usually from a pituitary tumor, that is, conditions involving cachexia or sarcopenia development. Its anabolic action also makes it attractive for a range of disorders with a catabolic (tissue breakdown) component, such as severe burns, cystic fibrosis, and inflammatory bowel disease, as well as in some fertility problems, osteoporosis, and Down's syndrome.[2,3]

Erythropoietin

In 1905, Paul Carnot and Clotilde Deflandre in France reported that plasma from anemic rabbits injected into normal rabbits caused an increase in erythrocyte (red blood cell) production, that is, erythropoiesis. They postulated that a single protein in the blood plasma caused this transformation, and they called it erythropoietin (Greek *erythro-*, red + *poiein*, to make). Erythropoiesis and hematopoiesis are sometimes confused and used interchangeably, but the latter is actually the development of all blood cells, including erythrocytes.[4,5]

It was not until the 1950s and 60s that several American investigators showed that stimulation of erythrocyte production was an endocrine hormonal effect, that the kidneys were the primary source of

EPO, and that low blood oxygen (hypoxemia) was the main stimulus for increased EPO production. In the early 1980s, methods for mass-producing synthetic EPO by recombinant DNA were developed, and the pharmaceutical company Amgen placed a highly successful version on the market.[4,5]

In recent years, alternative erythropoiesis-stimulating agents (ESAs), besides just recombinant human EPO (rhEPO), have emerged as viable PEDs and additional medical therapies (see Table 5.1). Various pharmaceutical versions of EPO and ESAs are used in severe anemia that develops from chronic kidney disease, inflammatory bowel disorders such as Crohn's disease and ulcerative colitis, and myelodysplasia resulting from of cancer chemotherapy and radiation.[4,5] Epoetin and Procrit are two of the most popular of the rhEPO forms currently prescribed. Table 5.1 is an overview of the most common ESAs used medically and as PEDs. Space limitation does not permit discussing all of these numerous agents.

INCREASING MUSCLE MASS AND ENHANCING OXYGEN DELIVERY

Growth Hormone

Endogenous growth hormone facilitates protein synthesis at the cellular level, which produces skeletal muscle accretion. This action in combination with exercise training leads to muscular hypertrophy and greater strength and power, obviously critical aspects for many sporting activities. The result is increased lean body mass (also called fat-free mass) and a reduced relative amount of stored body fat. This is the key reason athletes use growth hormone as a PED. Also, enhanced cellular protein synthesis produces a positive nitrogen balance nutritional state, which facilitates recovery from an exercise session and a training regime as well as being necessary for positive adaptation.[6] Growth hormone also works as a permissive and synergistic hormone to promote and amplify the actions of other anabolic hormones in the endocrine system such as insulin, testosterone, and insulin-like growth factors.[1,6]

Erythropoietin

Many sporting events demand a highly developed cardiovascular-respiratory system—heart, lungs, blood vessels, blood—to deliver oxygen to tissues. The more oxygen delivered to skeletal muscle, the more aerobic energy production in the form of adenosine triphosphate (ATP) can take place, which allows greater muscular work and improved performance, especially for endurance events.[6] A critical component in the delivery of oxygen is the oxygen-carrying capacity of the arterial blood—the oxygen content. Oxygen content is a function of the amount of hemoglobin (Hb) in the erythrocyte and the number of erythrocytes. Hb is the protein that binds, carries, and releases molecular oxygen. Since ESAs such as EPO stimulate erythropoiesis, increasing the number of erythrocytes and thus the amount of Hb, doping with EPO or other ESAs has a positive ergogenic effect on cardiovascular-respiratory capacity and thus potentially on aerobic, cardiovascular-respiratory performance.[6]

Growth hormone, EPO, and other ESAs have been on the World Anti-Doping Agency (WADA) banned substances list for several years. Nonetheless, they are highly popular PEDs. For example, National Football League player Bill Romanowski and baseball players Barry Bonds, Gary Sheffield, and Jason Giambi were accused of taking growth hormone in the controversial book *Game of Shadows*.[7] After a raid on the Bay Area Laboratory Co-Operative (BALCO) headquarters in 2003, owner Victor Conte claimed that he had supplied growth hormone to many high-profile American track athletes, including Marion Jones and Tim Montgomery. Jones, a multiple Olympic medal winner, admitted in 2007 to using PEDs, including growth hormone; she ultimately received a 6-month jail sentence for having previously denied it. Montgomery allegedly admitted to a US federal grand jury that he had taken growth hormone, and he received a 2-year ban. Conte's involvement in the BALCO scandal led to his imprisonment.[7] Disgraced professional cyclist Lance Armstrong admitted in his famous television interview with Oprah Winfrey that he took PEDs such as cortisone, EPO, growth hormone, and testosterone (and did blood doping) during all seven of his Tour de France victories. Blood doping is a general term indicating the use of PEDs or autologous or homologous transfusion of blood to boost oxygen-carrying capacity by increasing Hb.

RAMPING UP HORMONE ACTIONS WITHIN CELLS

Growth hormone and EPO have similar cellular mechanisms for invoking physiological changes, that is, secondary messenger hormonal signaling. This involves a cascade of molecular protein events to activate chemical reactions that bring about tissue responses. However, the aspects of how the stimuli produce the hormonal changes and resulting endocrine–paracrine–autocrine interactions with specific tissues differ between the two hormones.[1]

Growth Hormone

Growth hormone is highly anabolic and works directly and indirectly to bring changes at the cellular level. The direct effect is due to its ability to facilitate cellular uptake of glucose and the building blocks of protein (amino acids) from the blood. The indirect effects are via working in concert with growth factors, especially insulin-like growth factor 1 (IGF-1), to promote cellular protein synthesis and overall anabolic actions (Fig. 5.1). In the context of exercise training, the critical target tissue is skeletal muscle. Fig. 5.2 shows the interaction of growth hormone and the critical cellular mechanism by which IGF-1 activates the mammalian target of rapamycin complex 1 (see also Fig. 6.2), which is a key nonnuclear biochemical anabolic pathway for activating protein synthesis while inhibiting pathways of protein degradation.[8]

Erythropoietin

EPO specifically targets red bone marrow tissue and commences the differentiation process by which hemocytoblasts (stem cells) become reticulocytes and ultimately erythrocytes (Fig. 5.3). Specifically, EPO

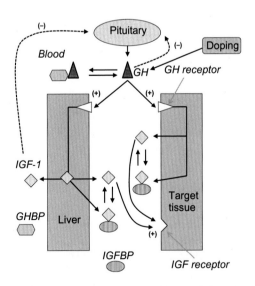

Figure 5.1 Regulation of growth hormone (GH) and insulin-like growth factor 1 (IGF-1). GHBP, growth hormone binding protein; IGFBP, IGF binding protein; solid arrows = stimulatory, dashed arrows = inhibitory.

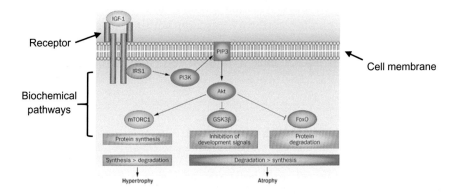

Figure 5.2 Binding of IGF-1 activates the PI3K—Akt pathway, leading to protein synthesis (anabolic action) and inhibition of the GSK3β pathway and FoxO degradation pathway. FoxO, forkhead box O protein; GSK3β, glycogen synthase kinase-3β; IGF-1, insulin-like growth factor I; IRS1, insulin receptor substrate 1; mTOR, serine/threonine-protein kinase mTOR; mTORC1, mTOR complex 1; PI3K, phosphoinositide-3-kinase; PIP3, phosphatidylinositol 3,4,5-trisphosphate. (Used with permission).[8]

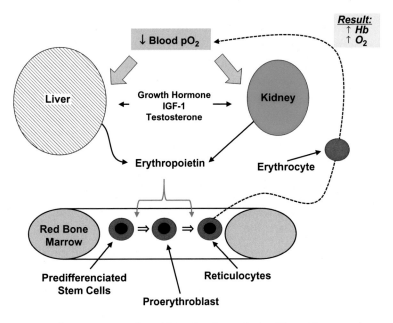

Figure 5.3 How erythropoietin increases hemoglobin and erythrocyte levels. pO2, partial pressure of oxygen; Hb, hemoglobin; O2, oxygen (capacity in blood); IGF-1, insulin-like growth factor 1; ↓ = decrease, ↑ = increase.

binds to its receptor on the hematopoietic stem cell and activates a Janus kinase 2 signaling pathway cascade within the cell. This biochemical processes results in differentiation, survival, and proliferation of the cells, which allows their evolution into mature erythrocytes in the blood stream.[9]

IT WORKS, WITHIN LIMITS

Growth Hormone

Since the late 1980s, growth hormone has been a popular ergogenic PED. Official and unofficial sources have reported a steady increase in misuse. Its attractiveness is based on the popular belief that it is efficient, hard to detect, and without major side effects if administered properly. These beliefs, however, are not well founded. The dose level and schedule of administration for optimum efficacy and safety are hard to determine, in part because athletes typically use a cocktail of PEDs concurrently (e.g., growth hormone, insulin, and testosterone). Underground information suggests that they take a supraphysiological dose of growth hormone three or four times a week in cycles of 4 to 6 weeks, as is the case with anabolic steroids used by some bodybuilders and strength and power athletes.[9–11] Anecdotal reports indicate dramatic increases in muscle mass and strength after large doses, but the effectiveness under controlled scientific conditions is generally far less impressive. The use of growth hormone in combination with agents such as EPO to enhance oxygen transport is gaining acceptance in endurance sport; here, the growth hormone is aimed more toward accelerating muscular recovery from exercise training and aiding erythropoiesis (Fig. 5.3).[9]

Growth hormone is also a potent lipolytic agent and results in stored body fat mobilization, triglyceride hydrolysis, and fatty acid utilization as an energy source.[10,11] The action occurs at the cellular and systemic levels by growth hormone synergistically increasing the production of other lipolytic hormones such as catecholamines and glucagon, and by increasing adipocyte adrenergic receptor expression.[12] When growth hormone is administered to people with growth hormone deficiency, body fat mass is reduced, and lean body mass is increased dramatically.[13] A meta-analysis of the research literature concluded that growth hormone in healthy athletes increases lean body mass, which is principally skeletal muscle, by an average of approximately 1.8 kg.[12]

Growth hormone induces glucose and amino acid uptake by the cell in various body tissues, thereby stimulating protein synthesis, an anabolic action.[11] This effect is supported by studies that show reduced muscle mass and protein synthesis when growth hormone and IGF-1 production are reduced or suppressed.[14,15] By contrast, growth hormone has little direct effect on protein degradation, but insulin and IGF-1 inhibit protein degradation (Fig. 5.2), and this may explain why some athletes use insulin and growth hormone concurrently.[10]

Growth hormone has an important function as an anabolic agent in skeletal muscle connective tissue, which is critical as it transmits force from individual muscle fibers to the bone. Strengthened connective tissue allows a stronger and more strain-resistant muscle and tendon unit, which allows great force development and resistance usage during exercise strength training.[16,17] Growth hormone also improves bone strength directly as well as indirectly through increased intestinal calcium absorption and maintenance of blood vitamin D levels.[18,19]

Erythropoietin

Investigations into the ergogenic effect of EPO (the rhEPO form) in aerobic-endurance sports occurred soon after it became readily available as a pharmaceutical.[20,21] High doses administered over 4 to 6 weeks result in substantial increases in maximal aerobic capacity (VO_{2max}) and exercise time to exhaustion.[20,22,23] In recent studies, physiological EPO doses increased VO_{2max} by 6% to12% (somewhat less than previously reported), hematocrit (Hct; relative erythrocyte count), and time to exhaustion while exercising at a given level of VO_{2max} (i.e., steady-state exercise intensity).[24–28] Why there is variance in the magnitude of VO_{2max} findings between studies is unclear but may relate to individual responsiveness differences (see the close-up in this chapter). EPO and ESAs are more effective when used in combination with dietary iron supplementation, as iron is a critical constituent in Hb.[28,29]

There are challenges in antidoping work with this PED. For example, research shows that when rhEPO administration is discontinued, VO_{2max} remains elevated for approximately 3 weeks even though blood EPO is reduced, decreasing the potential for detection.[27,28] Extended exposure to high altitude induces a natural increase in EPO which could mask doping use,[9] and altitude training is not banned by WADA. Also, several nonhematologic agents, such as hypoxia-inducible factor stabilizers (Table 5.1), improve aerobic-endurance performance by affecting in vivo EPO gene expression in the liver and kidney;[26–28] the athlete's blood EPO is endogenous and not of an exogenous source, which complicates doping detection.

Athlete Biological Passport and EPO

Standard bioanalytical detection methods (see Chapter 10; Athlete Testing, Analytical Procedures, and Adverse Analytical Findings) have major limitations for blood doping. Autologous blood reinfusion is difficult to detect, and with many ESAs the detection window is limited. Urine breakdown products of EPO are not highly specific. To address the problem, some international sports federations (cycling was the first) introduced upper acceptable limits for Hb and Hct in blood specimens. Athletes who tested above the limits were declared unfit for competition (no-start rule). However, Hb and Hct are influenced by many factors, such as body posture, high-altitude exposure, prior exercise, age, ethnicity, and gender.[9] And some nondoping athletes have naturally high values of these blood parameters.

These considerations led to the use of longitudinal blood profiles together with heterogeneous factors such as ethnicity and age to develop mathematical models to access doping likelihood. WADA adopted this concept in 2009 and developed the Athlete Biological Passport (ABP). The principle is to monitor selected hematological parameters over an extended time to indirectly reveal the effects of doping, rather than attempting to detect the doping substance or method itself. The ABP is complex and multifactorial and includes measurement of the following hematological outcomes from blood specimens: Hct, Hb, red blood cell count, reticulocyte percentage, reticulocyte number, mean corpuscular volume, mean corpuscular Hb mass, and mean corpuscular Hb concentration. Additional parameters of interest are the mean reticulocyte cell volume, reticulocyte Hb concentration, reticulocyte Hb content, serum EPO, and soluble transferrin receptors.[9] This technique is a complicated mathematical approach for monitoring potential dopers.

ACROMEGALY, BLOOD HYPERVISCOSITY, AND DEATH

Growth hormone and EPO used at PED doses induce serious side effects that can be life threating in some situations.

Growth Hormone

Growth hormone administered to a child or adult who does not have a hormonal deficiency can induce enlargement of the hands, feet, and face and abnormal growth of internal organs such as the heart, kidneys, and liver. Collectively these changes are called acromegaly. Diabetes, atherosclerosis, and hypertension may also be induced. A life-threatening effect of chronic use is cardiomyopathy secondary to acromegaly, characterized by significant worsening of the heart's ability to pump efficiently, which leads to heart failure if unchecked. There is also some evidence that high growth hormone levels are associated with cancerous tumor growth.[9] Less serious effects are neuromuscular and joint pain, edema, carpal tunnel syndrome, numbness or tingling of the skin, and elevated blood cholesterol.

Erythropoietin

Table 5.2 shows the major side effects of EPO and ESA. The most serious ones result from the blood becoming too viscous. Blood is not a pure fluid but rather a suspension of particles (cells) in a fluid (plasma). The use of EPO adds erythrocytes without the necessary concurrent increase in plasma volume; hence, the blood becomes thicker. A more viscous fluid encounters greater resistance to flow. In cardiovascular hemodynamics terms, this means the work of the heart increases and can exceed its capacity. Hyperviscosity worsens when the exercising athlete sweats, as a key source of the water for sweat is plasma.[6] Interestingly, there was an increase in suspicious deaths of professional cyclists during competitive events shortly after rhEPO became readily available. Increased viscosity and resultant strain on the heart is suspected as the cause.[9]

CONCLUSION

Most of the evidence supports the performance-enhancing properties of growth hormone, EPO, and related ESAs. The use of growth hormone as an adjunct to exercise training is a powerful anabolic stimulus that leads to muscle mass accretion. EPO and related ESAs are

Table 5.2 Side Effects of Erythropoietin When Used as a Performance-Enhancing Drug
Common and usually relatively mild
• Headache
• Body aches
• Diarrhea
• Cold symptoms (stuffy nose, sneezing, sore throat, cough)
• Joint pain
• Bone pain
• Muscle pain or spasms
• Dizziness
• Depression
• Weight loss
• Insomnia
• Nausea
• Vomiting
• Trouble swallowing
• Pain, tenderness, irritation (parenteral administration)
Serious
• Chest pain or heavy feeling, pain spreading to the arm or shoulder (possibly with nausea, sweating, general ill feeling)
• Feeling short of breath, lightheaded, or faint
• Swelling, rapid weight gain
• Sudden numbness or weakness (especially on one side of the body)
• Sudden severe headache, confusion, problems with vision, speech, or balance
• Pain, swelling, warmth, or redness in one or both legs
• Easy bruising, unusual bleeding from orifices
• Seizure (blackout or convulsions)
• Hypertension with severe headache, blurred vision, buzzing in the ears, anxiety, confusion, chest pain, irregular heartbeat

effective in improving the oxygen-carrying capacity of the blood, which can lead to improved performance. The efficacy of growth hormone and EPO as PEDs varies and is affected by dietary factors as well as individual sensitivity to these agents. Nonetheless, their use as PEDs remains prevalent and highly popular. They are banned, however, by antidoping agencies. Side effects can be severe, and some doping athletes have paid with their lives by using these PEDs.

Close-Up: Responders and Nonresponders—Does Blood Doping Always Work?

Martin Mooses
Faculty of Medicine, University of Tartu, Tartu, Estonia

Agents such as rhEPO and other ESAs are widely used in the treatment of anemia in clinical settings. So much so, that the introduction of rhEPO and ESAs has been described as a paradigm shift in the treatment of patients with clinical anemia due to, for example, chronic kidney disease, as it dramatically reduces the need for blood transfusions and their associated morbidity.[30,31] However, evidence indicates that hyporesponsiveness to rhEPO and ESAs occurs in approximately 5% to 10% of patients, creating an important diagnostic and management challenge in medical care.[30]

In sport sciences in recent years, there has been some speculation about whether hyporesponsiveness occurs in athletes. As early as 1998, Chapman et al. retrospectively investigated high-level and elite US distance runners' response to an altitude exposure of 2500 m (above sea level) and categorized those who improved after the 28 days of exposure as either responders or nonresponders to altitude training.[32] Responders were individuals who improved their 3000-m or 5000-m competition time trials (15 out of 32 participants) upon returning to sea level. An altitude of 2500 m induces a hypoxemia in the blood that should be a strong stimulus to EPO production (Fig. 5.3) and is an allowable "blood-doping" training procedure by WADA. These researchers showed that both responder and nonresponder groups demonstrated a significant increase in Hb concentration and Hct. However, after altitude exposure, responders had a significantly larger increase in mean blood EPO concentration, red cell volume, mean blood volume, and VO_{2max} compared with nonresponders.

Additionally, from the performance point of view, very recently, it was shown that rhEPO treatment enhanced performance in well-trained cyclists in a controlled laboratory-based maximal exercise test procedure that led to physical exhaustion but did not improve submaximal exercise testing responses or out-of-laboratory road race performance.[33]

These findings have caused some scientists to call into question whether all forms of blood doping really work and have provided the impetus for ongoing investigative work on these PEDs. Nonetheless, at this time, rhEPO and ESAs remain on the WADA banned list even if some researchers are calling into question their effectiveness.

REFERENCES

1. Constantini N, Hackney AC. *Endocrinology of Physical Activity and Sport*. New York: Humana Press; 2013.

2. Hintz R. Growth hormone: uses and abuses. *BMJ*. 2004;328(7445):907−908.

3. Menon PS. Growth hormone therapy: current status. *Indian Pediatr*. 1992;29(7):825−829.

4. Nissenson AR, Nimer SD, Wolcott DL. Recombinant human erythropoietin and renal anemia: molecular biology, clinical efficacy, and nervous system effects. *Ann Intern Med*. 1991;114(5):402−416.

5. Robinson N, Giraud S, Saudan C, et al. Erythropoietin and blood doping. *Br J Sports Med*. 2006;40(Suppl 1):i30−i34.

6. Stegemann J. *Exercise Physiology: Physiologic Bases of Work and Sport*. London, United Kingdom: Springer Verlag; 1981.

7. Fainaru-Wada M, Williams L. *Game of Shadows: Barry Bonds, BALCO, and the Steroids Scandal that Rocked Professional Sports*. New York: Gotham Books; 2006.

8. Hanaoka BY, Peterson CA, Horbinski C, et al. Implication of glucocorticoid therapy in idiopathic inflammatory myopathies. *Nat Rev Rheumatol*. 2012;8(8):448−457.

9. Jelkmann W, Lundby C. Blood doping and its detection. *Blood*. 2011;118(9):2395−2404.

10. Kraemer W, Nindl BC, Rubin MR. Growth hormone: physiological effects of exogenous administration. In: Bahrke MS, Yesalis CE, eds. *Performance Enhancing Substances in Sport and Exercise*. Champaign, IL: Human Kinetics; 2002:65−78.

11. Sonksen PH. Insulin, growth hormone and sport. *J Endocrinol*. 2001;170:13−25.

12. Mekala KC, Tritos NA. Effects of recombinant human growth hormone therapy in obesity in adults: a meta analysis. *J Clin Endocrinol Metab*. 2009;94:130−137.

13. Binnerts A, Swart GR, Wilson JH, et al. The effect of growth hormone administration in growth hormone deficient adults on bone, protein, carbohydrate and lipid homeostasis, as well as on body composition. *Clin Endocrinol (Oxf)*. 1992;37:79−87.

14. Fryburg DA, Gelfand RA, Barrett EJ. Growth hormone acutely stimulates forearm muscle protein synthesis in normal humans. *Am J Physiol*. 1991;260:E499−E504.

15. Fryburg DA, Barrett EJ. Growth hormone acutely stimulates skeletal muscle but not whole-body protein synthesis in humans. *Metabolism*. 1993;42:1223−1227.

16. Rennie MJ. Claims for the anabolic effects of growth hormone: a case of the emperor's new clothes? *Br J Sports Med*. 2003;37:100−105.

17. Doessing S, Kjaer M. Growth hormone and connective tissue in exercise. *Scand J Med Sci Sports*. 2005;15:202−210.

18. Lund B, Eskildsen PC, Lund B, et al. Calcium and vitamin D metabolism in acromegaly. *Acta Endocrinol (Copenh)*. 1981;96:444−450.

19. Burstein S, Chen IW, Tsang RC. Effects of growth hormone replacement therapy on 1,25-dihydroxyvitamin D and calcium metabolism. *J Clin Endocrinol Metab*. 1983;56:1246−1251.

20. Ekblom B, Berglund B. Effect of erythropoietin administration on maximal aerobic power in man. *Scand J Med Sci Sports*. 1991;1(2):88−93.

21. Berglund B, Ekblom B. Effect of recombinant human erythropoietin treatment on blood pressure and some haematological parameters in healthy men. *J Intern Med*. 1991;229(2):125−130.

22. Audran M, Gareau R, Matecki S, et al. Effects of erythropoietin administration in training athletes and possible indirect detection in doping control. *Med Sci Sports Exerc*. 1999;31(5):639−645.

23. Birkeland KI, Stray-Gundersen J, Hemmersbach P, et al. Effect of rhEPO administration on serum levels of sTfR and cycling performance. *Med Sci Sports Exerc.* 2000;32(7):1238—1243.

24. Thomsen JJ, Rentsch RL, Robach P, et al. Prolonged administration of recombinant human erythropoietin increases submaximal performance more than maximal aerobic capacity. *Eur J Appl Physiol.* 2007;101(4):481—486.

25. Lundby C, Olsen NV. Effects of recombinant human erythropoietin in normal humans. *J Physiol.* 2011;589(6):1265—1271.

26. Lundby C, Robach P, Boushel R, et al. Does recombinant human EPO increase exercise capacity by means other than augmenting oxygen transport? *J Appl Physiol.* 2008;105(2):581—587.

27. Lundby C, Hellsten Y, Jensen MB, et al. Erythropoietin receptor in human skeletal muscle and the effects of acute and long-term injections with recombinant human erythropoietin on the skeletal muscle. *J Appl Physiol.* 2008;104(4):1154—1160.

28. Rasmussen P, Foged EM, Krogh-Madsen R, et al. Effects of erythropoietin administration on cerebral metabolism and exercise capacity in men. *J Appl Physiol.* 2010;109(2):476—483.

29. Major A, Mathez-Loic F, Rohling R, et al. The effect of intravenous iron on the reticulocyte response to recombinant human erythropoietin. *Br J Haematol.* 1997;98(2):292—294.

30. Johnson DW, Pollock CA, Macdougall IC. Erythropoiesis-stimulating agent hyporesponsiveness. *Nephrology.* 2007;12(4):321—330.

31. Durussel J, Haile DW, Mooses K, et al. Blood transcriptional signature of recombinant human erythropoietin administration and implications for antidoping strategies. *Physiol Genomics.* 2016;48(3):202—209.

32. Chapman RF, Stray-Gundersen J, Levine BD. Individual variation in response to altitude training. *J Appl Physiol.* 1998;85(4):1448—1456.

33. Heuberger JAAC, Rotmans JI, Gal P, et al. Effects of erythropoietin on cycling performance of well-trained cyclists: a double-blind, randomised, placebo-controlled trial. *Lancet Haematol.* 2017;S2352—3026(17):30105—30109.

Beta-2 Agonists

Beta-2 agonists are an important group of drugs that selectively mimic the actions of the endogenous catecholamines epinephrine and norepinephrine, two major neurotransmitters-hormones. As pharmaceuticals, the major use of beta-2 agonists is to reduce signs and symptoms of asthma and chronic obstructive pulmonary disease (COPD) by bronchodilation, allowing the patient to breathe more easily. Their medical effectiveness is well documented, and they are widely prescribed. As performance-enhancing drugs, they increase pulmonary function and have many downstream actions on the cardiovascular and metabolic systems. One such action is activation of nonsteroidal anabolic processes leading to muscle mass enhancement.

ASTHMA PREVENTION AND TREATMENT

The Japanese scientist Dr. Jokichi Takamine isolated the hormone epinephrine around 1900, and by the 1930s it was in use therapeutically for asthma.[1] Although viewed as a medical breakthrough, it had adverse side effects such as anxiety, restlessness, headache, dizziness, and heart palpitations. In the 1940s, the catecholamine-like drug isoproterenol was discovered and determined to have the beneficial effects of epinephrine with fewer and milder side effects. This was the first beta-2-agonist and was unselective in its action (mechanism of action is explained in the following section). It was termed as sympathomimetic because of its epinephrine-like effects; norepinephrine and epinephrine are associated with the sympathetic division of the autonomic nervous system as both neurotransmitters and hormones.[5,9] By the beginning of the 1950s, isoproterenol was in general used for asthma and other lung conditions.[1] Initially, the route of administration was sublingual or by inhalation using a squeeze-bulb. The first pressurized metered-dose inhaler was introduced in the mid-1950s, and it was far more efficient. By the 1970s, inhaler technology was fully developed to the level we recognize today.

Doping, Performance-Enhancing Drugs, and Hormones in Sport. DOI: https://doi.org/10.1016/B978-0-12-813442-9.00006-7

Research in the 1940s and 1950s revealed that many tissues in the body had adrenergic receptors of several subtypes that responded to catecholamine neurotransmitters. In the 1960s, activation of the beta-2 adrenergic receptor was shown to be the mechanism responsible for the bronchodilation and asthmatic relief provided by sympathomimetics. This discovery encouraged pharmaceutical companies to seek more selective drugs to aid patients with asthma and COPD.[2] Orciprenaline, a longer-acting beta-2 agonist, was discovered, followed in quick succession by albuterol, salbutamol, terbutaline, and fenoterol (see Table 6.1 for commonly prescribed beta-2 agonists). Most of these newer drugs had fewer side effects than isoproterenol, as they were more selective for beta-2 receptors. In the 1990s, salmeterol, a highly effective and long-acting beta-2 agonist bronchodilator, was marketed, followed by formoterol. In 2013, vilanterol, an even longer-lasting bronchodilator, was marketed; its effects last approximately 24 hours.[1]

Asthma drugs are categorized as quick relief (reliever medications) and long-term control (controller medications). The former is used to relieve acute exacerbations and prevent exercise-induced

Table 6.1 Commonly Prescribed Beta-2 Agonists	
Generic Name	Brand or Trade Name
Short- and Long-Acting Beta-2 Agonists	
albuterol (short-acting)	Proair, Proventil, Ventolin
arformoterol (long-acting)	Brovana
formoterol (long-acting)	Foradil, Perforomist
indacaterol (long-acting)	Arcapta Neohaler
levalbuterol (short-acting)	Xopenex
metaproterenol (short-acting)	Alupent
salmeterol (long-acting)	Serevent
terbutaline (short-acting)	Brethine
Long-Acting Beta-2 Agonist and Anticholinergic Combination	
vilanterol, umeclidinium	Anoro Ellipta, Incruse Ellipta
Long-Acting Beta-2 Agonist and Corticosteroid Combination	
formoterol, budesonide	Symbicort
salmeterol, fluticasone	Advair
vilanterol, fluticasone	Breo Ellipta
Short-Acting Beta-2 Agonist and Anticholinergic Combination	
albuterol, ipratropium	Combivent, DuoNeb

bronchoconstriction; the devices are sometimes called rescue inhalers. They include short-acting beta agonists, anticholinergics (used only in severe cases), and systemic corticosteroids, which speed recovery from acute problems. Long-term controller agents are primarily comprised of inhaled corticosteroids (glucocorticoid-related drugs), long-acting beta agonists, long-acting anticholinergics, and long-acting beta-2 agonists in combination with anticholinergics or corticosteroids (see Chapter 4: Glucocorticoids for explanation of their physiological actions and roles). Inhaled corticosteroids are drugs of choice for control of chronic asthma, though efficacy varies widely among patients.[2]

BREATHING MORE AND BUILDING MUSCLE IN SPORT

Beta-2 agonists are potentially powerful doping agents because they stimulate receptors in the central and peripheral nervous systems as well as tissues that control a range of exercise functions, among them bronchodilation, anabolic processes, and enhancement of the anti-inflammatory effects of corticosteroids.[2–4]

In exercise, bronchodilation is of benefit in cardiovascular-respiratory-based aerobic sports, as it increases the amount of air moved in and out of the body. Such sporting activities rely greatly on the body's ability to deliver oxygen to the exercising muscle for use in aerobic energy metabolism, i.e., adenosine triphosphate (ATP) production.[5] The components of oxygen delivery are linked to the ability of the body to carry oxygen in the blood (oxygen content) and the ability of the lungs to move oxygen from the atmosphere into the blood (external respiration).[6] Improvement of either oxygen content or external respiration can dramatically improve oxygen delivery to skeletal muscle and thus facilitate exercise performance (see Chapter 5: Peptide-Protein Hormones).[5,6] While these points are physiologically sound, it is important to note that the ability to move air in and out of the lungs (pulmonary function) is seldom a limiting factor in exercise for healthy individuals (see the close-up at the end of this chapter). Nonetheless, bronchodilation can be seen as a performance enhancer, which is cited as a rationale for beta-2 agonist use by some doping athletes.

Perhaps, less obvious yet important is that some beta-2 agonist formulations have nonsteroidal anabolic action that increases muscle mass, which is beneficial to weightlifters and other strength-power athletes.[7] The mechanism involves cellular cytosolic anabolic biochemical pathways, which are explained in the following sections.

The World Anti-Doping Agency (WADA) prohibits the use of beta-2 agonists, even in asthmatics, without a therapeutic use exemption unless the standard of care medically dictates their use (e.g., in rescue inhaler situations). In some cases, a drug is permitted subject to daily use restrictions. Strict medical records and physician documentation of what happened to require usage, why it happened, and how much drug was administered must be kept.[8]

ADRENERGIC AND ANABOLIC ACTIONS AND REACTIONS

To understand the mechanism of action of beta-2 agonists requires understanding how adrenergic receptors function. Epinephrine and norepinephrine are both neurotransmitters and hormones that activate adrenergic receptors. They are secreted by the sympathetic nervous system and adrenal medulla gland into the bloodstream and are increased at times of emotional or physical stress to facilitate a series of physiological responses—a stress reactivity coping mechanism.[9] In an exercise session and especially in competitive sports events, the result is increased cardiac output (volume of blood pumped by the heart), pulmonary ventilation (air moved by the lungs), increased circulating glucose from the liver (to use as an energy source in ATP metabolism), and dilation of blood vessels in skeletal muscle to allow increased blood flow and oxygen delivery. Secretion and circulation of epinephrine and norepinephrine are proportional to the intensity of activity, but the response is curvilinear rather than linear.[6,9]

Epinephrine and norepinephrine bring about their physiological effects by binding to alpha-1, beta-1, alpha-2, and beta-2 adrenergic subtypes in a variety of tissues.[10] Fig. 6.1 shows some of the wide-ranging implications of adrenergic receptor activation.

Current beta-2 agonists work by selectively activating only beta-2 receptors. Fig. 6.2 shows the cellular mechanisms of action. Once the receptor is activated, there is a cascade of biochemical reactions in which metabolic pathways are turn-on or turn-off. For bronchodilation, receptor activation results in relaxation of the smooth muscle of the air passages, allowing enlargement of the bronchial tissues and reduction of the resistance to air flow. For skeletal muscle, receptor

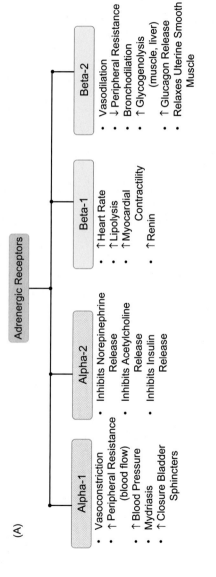

Figure 6.1 Adrenergic receptor subtypes. (A) Physiological effects of catecholamine binding to the specific receptor. (B) Affinity of the receptor subtypes for each catecholamine. ↑ = increase, ↓ = decrease.

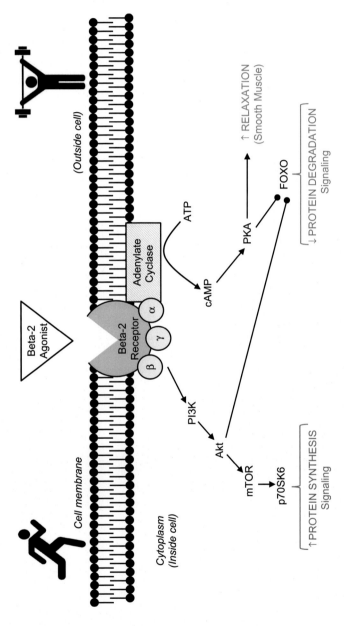

Figure 6.2 The nonsteroidal biochemical pathway for induction of protein synthesis in cells. Arrows indicate activation of a mechanism, lines with • indicate inhibition. Abbreviations: β, γ, α, receptor subprotein components; PI3K, phosphatidylinositol 3-kinase; Akt, protein kinase B; mTOR, mammalian target of rapamycin; p70SK6, ribosomal protein S6 kinase beta-1; ATP, adenosine triphosphate; cAMP, cyclic adenosine monophosphate; PKA, protein kinase A; FOXO, Forkhead box proteins. ↑ = increase, ↓ = decrease.

activation results in nonsteroidal cytoplasmic PI3K/Akt/mTOR/ p70SK6 anabolic pathway activation and FOXO catabolic pathway inhibition. These actions shift the protein turnover balance of the cells toward greater synthesis and less degradation.[5–9,11]

ASTHMATICS, YES OR MAYBE! EVERYONE ELSE?

Research findings on the exercise performance benefits of beta-2 agonists via improved bronchodilation in nonasthmatic athletes appear contradictory.[2,4] Some researchers found no improvement in aerobic capacity (maximum amount of oxygen that can be used at maximum effort, VO_{2max}), performance in endurance events, or peak cycling power output,[4,11] and no effect on oxygen delivery kinetics at the tissue level.[12] On the other hand, Pluim et al. reported some benefits in endurance performance and cycling sprint power output with very high doses acutely or chronically administered,[11] and Hostrup et al. found a ~4% improvement in sprint power output of cyclists with a similar administration protocol.[13] In spite of these findings, the Joint Task Force of the European Respiratory Society and European Academy of Allergy and Clinical Immunology emphatically states that there is no ergogenic benefit of air flow improvement in nonasthmatics, but anecdotal evidence suggest that athletes still use beta-2 agonists in this capacity, with the combination of salbutamol and albuterol being popular.[14,15]

There is strong evidence of improved muscular strength and endurance in resistance training.[4,15,16] The nonsteroidal anabolic agent clenbuterol is specifically used in this capacity (see Fig. 6.2 for mechanism of action). Its use increases skeletal muscle contractile proteins and muscle fiber mass, which explains its preferential use by weightlifters and other strength-power athletes.[7,15–17] It is also used in the livestock industry for these anabolic effects, and some reports suggest that eating the meat of animals treated with it can result in a positive doping test.[18] For example, in 2010, clenbuterol was found in a urine sample of the Spanish professional cyclist Alberto Contador, who had just won the Tour de France. He used the argument of "contaminated meat" as an explanation for the positive doping result, but after a thorough review by antidoping agencies his victory was revoked and he was banned for 2 years.[15]

TACHYCARDIA, ARRHYTHMIAS, SYNCOPE

Because beta-2 agonists bind to their receptors on many types of tissue (Fig. 6.1), they have many side effects. Excessive activation on the heart involves the greatest health risk, potentially causing arrhythmias, palpitations, and myocardial ischemia (reduced blood flow to the blood vessels of the heart, potentially precipitating a heart attack), all of which can be life threating.[3,4,19]

Less severe but commonly encountered side effects are

- headache
- anxiety
- nausea and dizziness
- syncope
- nervousness
- sweating
- insomnia

There are also reports of muscle tremor (sometime severe) and increased circulating blood glucose concentration due to drug action on the liver. Excessive glucose can lead to polydipsia (increased thirst) because it increases the osmolality of blood, making it more concentrated, which is a signal to the thirst center in the hypothalamus of the brain. Excessive blood glucose can also lead to polyuria (increased frequency of urination) from the increased fluid intake and increased urine formation at the kidney. The overall result is dehydration, which compromises exercise capacity and performance (see Fig. 10.1).[5,6] There is also a slight risk of glucose-related kidney and blood vessel damage from the hyperglycemia.[19]

A side effect of clenbuterol is reduction in bone mineral content,[7] which increases the risk of osteopenia and osteoporosis, leading to stress fractures and ultimately a curtailment of exercise training.[5,6]

CONCLUSION

In the treatment of bronchospasm associated with asthma, beta-2 agonists are critical and a medical mainstay. Their development was a scientific breakthrough. Even though they are in widespread medical use, they are banned by antidoping agencies such as WADA. The rationale is that they are associated with enhanced acute cardiovascular-respiratory-based performance and facilitate aspects of physiological adaptation in response to exercise training, especially in skeletal muscle. This latter role as an anabolic agent accounts for the most prevalent and persistent use by doping athletes. The capacity of these drugs to aid asthmatic athletes is well documented; but use by such individuals still requires a therapeutic use exemption.

Close-Up: Use of Inhalers by Asthmatic Athletes— Is This Really Doping?

By Hans Christian Haverkamp
Johnson State College, Johnson, VT, United States

Joseph W. Duke
Northern Arizona University, Flagstaff, AZ, United States

Airway narrowing is the principal physiologic abnormality in bronchial asthma and causes the well-known symptoms of breathing discomfort, chest tightness, and cough. Given that whole-body exercise is one of the most—if not the most—potent triggers of airway narrowing in asthma, it is not surprising that use of inhaled β2-agonists is ubiquitous in asthmatic athletes. This widespread use, however, begs the question by the scientists: Do inhaled β2-agonists improve exercise performance in asthmatic athletes, and if so, by what mechanism?

Inhaled β2-agonists do not improve endurance performance in nonasthmatic athletes.[14] Similarly, they do not reliably improve endurance performance in asthmatic persons.[20,21] This lack of ergogenicity in asthmatics is perhaps surprising not only due to the bronchodilation (BD)-induced increase in airway diameter but also to the efficacy of β2-agonists for preventing postexercise airway narrowing (i.e., exercise-induced asthma). Why, then, are inhaled β2-agonists apparently so ineffective at improving exercise performance in the asthmatic?

A potent exercise-induced BD is likely the most important reason for the limited impact of inhaled β2-agonists on exercise performance in asthmatics. In both nonasthmatic and asthmatic persons, whole-body exercise causes BD that is rapid and exercise intensity dependent.[22–24] In the asthmatic with narrowed airways at baseline, the amount of exercise-induced BD can be considerable. Fig. 6.3A shows a substantial increase in the maximal expiratory flow−volume curve immediately after cycling exercise compared with before the exercise in one asthmatic individual.

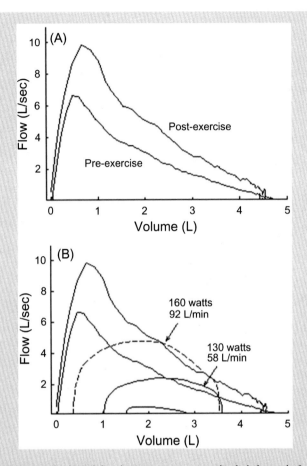

Figure 6.3 (A) Maximal volitional forced expiratory curves immediately before and after exercise in one asthmatic subject. (B) Spontaneous exercise tidal flow—volume curves at two exercise workloads (i.e., watts) plotted within the maximal expiratory curves measured pre- and postexercise.

This demonstrates significant exercise-induced BD. In Fig. 6.3B, tidal exercise flow—volume curves are plotted within the maximal curves. Notice that exercise airflow at both workloads exceeds the preexercise maximal airflow but not the postexercise maximal airflow. Thus, the exercise ventilation rates would not have been possible without the BD that occurred during the exercise. Undoubtedly, if the BD had not occurred, the ventilatory mechanical constraint would have limited exercise capacity in this subject. Thus, in many instances, exercise-induced BD will obviate the potential for dilated airways to improve exercise performance.

Despite the exercise BD, mechanical constraints to ventilation can occur when the requirements for airflow reach the limits imposed by the structural and mechanical properties of the airways.[25] Moreover, in the asthmatic with fixed airway narrowing that is not responsive to β2-agonists, exercise ventilatory limitation will be more likely to occur. If ventilatory constraint does occur, it will be one among the various mechanisms limiting exercise capacity. Indeed, improved airway function due to treatment with inhaled steroid drugs (to treat airway inflammation) resulted in significant increases in ventilation and exercise performance in a group of asthmatics.[26]

Is it possible to predict if preexercise β2-agonists will improve exercise performance in an individual asthmatic? The short answer is "No." First, airway function varies by day, week, month, and year within an asthmatic.[27] Thus, the propensity for reaching exercise ventilatory limitation will also be variable. Second, the extent of exercise-induced BD is variable among asthmatics. Much of this variability relates not only to airway function at baseline but also to whether any "fixed" airway narrowing has occurred. Finally, the exercise ventilatory requirement—which ultimately depends on aerobic capacity—will interact with the first two variables to determine whether ventilatory constraint is reached.

All of the above said, perhaps the best answer to the initial question asked is "maybe" inhaled β2-agonists can improve exercise performance in some asthmatic athletes, depending on the interactions among the many factors influencing the response. That said, the WADA maintains β2-agonists on it banned list, and to avoid testing positive, asthmatic athletes must apply for a therapeutic use exemption.

REFERENCES

1. Tattersfield AE. Current issues with beta-2-adrenoceptor agonists: historical background. *Am J Med.* 2006;68(4):471−472.

2. Celli BR, MacNee W. "Standards for the diagnosis and treatment of patients with COPD: a summary of the ATS/ERS position paper". *Eur Respir J.* 2006;27(1):242.

3. Kindermann W. Do inhaled beta(2)-agonists have an ergogenic potential in non-asthmatic competitive athletes? *Sports Med.* 2007;37(2):95−102.

4. Kindermann W, Meyer T. Inhaled β2 agonists and performance in competitive athletes. *Br J Sports Med.* 2006;40(Suppl 1):i43−i47.

5. Hackney AC. *Exercise, Sport, and Bioanalytical Chemistry: Principles and Practice.* New York, New York: Elsevier−RTI Press; 2016.

6. Stegemann J. *Exercise Physiology: Physiologic Bases of Work and Sport.* London: Springer; 1981.

7. Douillard A. Skeletal and cardiac muscle ergogenics and side effects of clenbuterol treatment. *J Sport Med Doping Stud.* 2011;S1:001.

8. USADA. Coaches: how to help athletes with inhalers avoid doping violations. <https://www.usada.org/coaches-how-to-help-athletes-with-inhalers-avoid-doping-violations/>; 2016 Accessed 17.07.28.

9. Hackney AC. Stress and the neuroendocrine system: the role of exercise as a stressor and modifier of stress. *Expert Rev Endocrinol Metab*. 2006;1(6):783−792.

10. Davis E, Loiacono R, Summers RJ. The rush to adrenaline: drugs in sport acting on the b-adrenergic system. *Br J Pharmacol*. 2008;154:584−597.

11. Pluim BM, de Hon O, Staal JB, et al. ß2-agonists and physical performance: a systematic review and meta analysis of randomized controlled trials. *Sports Med*. 2011;41:39−57.

12. Elers J, Mørkeberg J, Jansen T, et al. High-dose inhaled salbutamol has no acute effects on aerobic capacity or oxygen uptake kinetics in healthy trained men. *Scand J Med Sci Sports*. 2012;22:232−239.

13. Hostrup M, Kalsen A, Auchenberg M, Bangsbo J, Backer V. Effects of acute and 2-week administration of oral salbutamol on exercise performance and muscle strength in athletes. *Scand J Med Sci Sports*. 2016;26(1):8−16.

14. Carlsen KH, Anderson SD, et al. Treatment of exercise-induced asthma, respiratory and allergic disorders in sports and the relationship to doping: Part II of the report from the Joint Task Force of European Respiratory Society (ERS) and European Academy of Allergy and Clinical Immunology (EAACI) in cooperation with GA(2)LEN. *Allergy*. 2008;63(5):492−505.

15. Bird RB, Goebel C, Burke LM, et al. Doping in sport and exercise: anabolic, ergogenic, health and clinical issues. *Ann Clin Biochem*. 2016;53(2):196−221.

16. Caruso JF, McLagan JR, Olson NM, et al. Beta(2)-adrenergic agonist administration and strength training. *Phys Sportsmed*. 2009;37:66−73.

17. West DWD, Phillips SM. Anabolic processes in human skeletal muscle: restoring the identities of growth hormone and testosterone. *Phys Sportsmed*. 2010;38(3):97−104.

18. Guddat S, Fußholler G, Geyer H, et al. Clenbuterol—regional food contamination a possible source for inadvertent doping in sports. *Drug Test Anal*. 2012;4:534−538.

19. Huckins DS, Lemons MF. Myocardial ischemia associated with clenbuterol abuse: report of two case studies. *J Emergency Med*. 2013;44:444−449.

20. Freeman W, Packe GE, Clayton RM. Effect of nebulised salbutamol on maximal exercise performance in men with mild asthma. *Thorax*. 1989;44(11):942−947.

21. Robertson W, Simkins J, O'Hickey SP, et al. Does single dose salmeterol affect exercise capacity in asthmatic men?" *Eur Respir J*. 1994;7(11):1978−1984.

22. Haverkamp HC, Dempsey JA, Miller JD, et al. Gas exchange during exercise in habitually active asthmatic subjects. *J Appl Physiol*. 2005;99(5):1938−1950.

23. Milanese M, Saporiti R, Bartolini S, et al. Bronchodilator effects of exercise hyperpnea and albuterol in mild-to-moderate asthma. *J Appl Physiol*. 2009;107(2):494−499.

24. Klansky A, Irvin C, Morrison-Taylor A, et al. No effect of elevated operating lung volumes on airway function during variable workrate exercise in asthmatic humans. *J Appl Physiol*. 2016;121(1):89−100.

25. Haverkamp HC, Dempsey JA, Miller JD, et al. Repeat exercise normalizes the gas-exchange impairment induced by a previous exercise bout in asthmatic subjects. *J Appl Physiol*. 2005;99(5):1843−1852.

26. Haverkamp HC, Dempsey JA, Pegelow DF, et al. Treatment of airway inflammation improves exercise pulmonary gas exchange and performance in asthmatic subjects. *J Allergy Clin Immunol*. 2007;120(1):39−47.

27. U.S. Department of Health and Human Services, *National Heart, Lung and Blood Institute*. Expert panel report 3: guidelines for the diagnosis and management of asthma, NIH Publication Number 08-584.2007.

CHAPTER 7

Hormone and Metabolic Modulators

Selective androgen receptor modulator (SARM) and Selective estrogen receptor modulator (SERM) drugs are relatively new doping agents. They vary in type according to physiological intent, but the general mechanism of action is uniform: They affect cellular receptors in an endocrine-like action on tissue that is responsible for implementing positive changes in the cell environment. These changes are down-stream events. For athletes, the use of these agents can mean enhanced protein synthesis and anabolic actions at skeletal muscle, leading to functional and performance enhancement.

ANDROGEN AND ESTROGEN MODULATION IN HEALTH AND DISEASE

A selective receptor modulator drug is one that influences specific hor-monal receptors on a tissue. In the simplest terms, the drug can have an agonistic or antagonistic effect on the receptor, either activating or blocking it from becoming active (Fig. 7.1). Some drugs are mixed

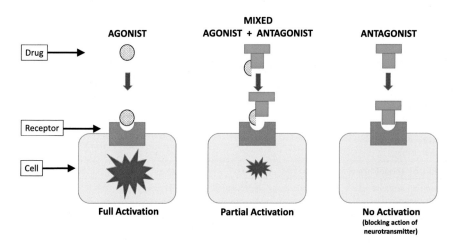

Figure 7.1 Actions of agonist and antagonist drugs on cellular receptors.

Doping, Performance-Enhancing Drugs, and Hormones in Sport. DOI: https://doi.org/10.1016/B978-0-12-813442-9.00007-9

agonist—antagonists. Tissue-selective drugs stand in contrast to those that produce global agonism or antagonism, wide-ranging actions that can result in the desired effect but also undesirable side effects.[1] The pharmaceutical industry produces three categories of these drugs (Table 7.1):

- SARMs
- SERMs
- Selective progesterone receptor modulators (SPRMs)

SARMs are intended to have the same effects as androgenic—anabolic drugs such as steroids and hormones such as testosterone, but to do so more selectively. This allows them to be used for more clinical indications than the relatively limited uses for which anabolic steroids are currently approved. SARMs are designed to differentiate between anabolic and androgenic physiologic activation, thus lending themselves to use in a variety of medical conditions such as muscle-wasting diseases, osteoporosis, cancer, and hypogonadism.[1,2] They can be delivered

Table 7.1 Examples of and Medical Uses for SARMs, SERMs, and SPRMs	
Examples	**Treatment Uses[a]**
Selective Androgen Receptor Modulators (SARMs)	
enobosarm (Ostarine) BMS-564,929 LGD-4033 (Ligandrol) andarine (S4) RAD140 (Testolone) MK-0773 S-22 YK11	Hypogonadism (males), osteopenia, osteoporosis, sarcopenia, cachexia, andropause
Selective Estrogen Receptor Modulators (SERMs)	
clomiphene (Clomid) ormeloxifene raloxifene tamoxifen toremifene lasofoxifene ospemifene	Breast cancer, infertility, osteoporosis, vaginal atrophy, dyspareunia, contraception, male hypogonadism, gynecomastia, breast pain
Selective Progesterone Receptor Modulators (SPRMs)	
ulipristal acetate ("Ella") asoprisnil (J867) telapristone (CDB-4124) proellex	Emergency contraception, uterine fibroids

[a]There are many SARMS, SERMS, and SPRMs that have specific medical uses. Not all of those listed have all treatment uses listed.

orally; many other anabolic-like drugs are only injectable. The objective is a customized drug response; i.e., the target tissues will respond as they would to testosterone, and nontarget tissues will not respond, thereby avoiding side effects. Existing SARMs are not as selective for muscle and bone as desired. For example, they produce anabolic–androgenic effects in the prostate gland, increasing the risk of prostate cancer. Nonetheless, they have a greatly reduced ratio of anabolic to androgenic effects compared to the endogenous hormone testosterone.[3]

SERMs act on estrogen receptors, which occur in many types of tissue.[4] As with SARMs, they differ from pure estrogen receptor agonists and antagonists in that their action is different in different tissues, resulting in selective stimulation of estrogen-like action. Tissue types have different degrees of sensitivity and thus reactivity. So a SERM can produce estrogenic or antiestrogenic effects depending on the specific tissue as well as the percentage of intrinsic activity of the drug.[4] Intrinsic activity is the relative ability of a drug-receptor complex to produce a maximum functional physiological response (*N. B.*: This concept is applicable to all drugs that work by affecting cell receptors).[5] This ability is different from drug-receptor affinity, which is a measure of the ability of the drug to bind to its molecular target. SERMs with high intrinsic activity have more estrogenic effects, while those with low intrinsic activity have primarily antiestrogenic effects. Two of the most commonly used SERMs are clomiphene and tamoxifen, which have more of an equal balance in estrogenic and antiestrogenic actions.

SPRMs follow the same principles of action and effect as SARMs and SERMs, but on receptors of the reproductive hormone progesterone. SPRMs are not associated with sport doping and therefore seem not to be used as performance-enhancing drugs (PEDs). They are not discussed further in this chapter.

THE BALANCE BETWEEN GENDER-BASED HORMONES

SARMs have properties similar to anabolic agents but with less androgenic action. Thus, they have the advantage of androgen-receptor specificity, tissue selectivity, and few steroid-related side effects. These qualities make SARMs preferred by doping athletes, who want the anabolic action without the acne, liver damage, breast tissue development, and shrinkage of testicles in males, the deepened voice, hair growth on face, stomach, and upper back, and abnormal menstrual cycles in females. Chapter 2: Anabolic Androgenic Steroids explains the anabolic actions of testosterone and steroids, which SARMs mimic, and why sporting athletes benefit physiologically from such agents.[3]

SARMs have been prohibited by the World Anti-Doping Agency (WADA) since 2008 because they are assumed to enhance performance by stimulating androgen receptors in muscle and bone (i.e., in the same fashion as testosterone or anabolic–androgenic steroids). They are prohibited at all times and are in the category of "other anabolic agents" under Section S1.2 of the WADA Prohibited List (see Chapter 1: Overview: Doping in Sport). Ostarine and andarine are among the most popular SARMs currently used as PEDs by athletes.[6]

SERMs are used as PEDs because of their antiestrogenic action, most often by men. In women, they act on the pituitary gland to stimulate the release of specific hormones responsible for ovulation. In men, they alter testosterone levels by interfering with the negative feedback loop of the hypothalamic–pituitary–gonadal axis, thereby allowing testosterone levels to rise much higher than normal.[7]

The number of athletes testing positive for SARMs and SERMs has grown over the last decade despite inclusion on the Prohibited List.[6] A recent high-profile case involved Jon Jones, a mixed martial arts fighter, who tested positive for two banned substances, clomiphene and letrozole. In the doping world, clomiphene is a postcycle therapy drug, that is, something used after using anabolic steroids to raise endogenous testosterone levels. Letrozole is an aromatase inhibitor. Aromatase is the cellular enzyme that converts testosterone into estrogen, that is, peripheral aromatization. Inhibiting the aromatase enzyme leaves more natural testosterone in the system. Clinically, letrozole is used to treat breast cancer. In men, it works in much the same way as clomiphene but has fewer negative side effects.[2,7]

SARMs are not legal ingredients for dietary supplements, but they have nonetheless been detected in such products sold illegally, and they could pose significant health risks to athletes. Furthermore, since SARMs are on the WADA List, consumption of the products even unknowingly or accidently could lead to an antidoping rule violation. As an illustration, in 2016, the American triathlete Lauren Barnett was given a suspension for using ostarine. She was able to establish that it came from a contaminated supplement and so was given a reduced ban (only 6 months). Similar situations have occurred in many athletes.[2]

Athletes are not the only ones receiving bans for inappropriate actions. Recently, Michael Gingras, a coach in the sport of weightlifting, received a 12-year sanction for providing athletes with a variety of doping agents, one of which was a SARM.[2]

The US Anti-Doping Agency recommends that athletes be aware that SARM ingredients could be listed on dietary supplement product labels under various names, and not fully identified as SARMs.[8] In 2014, the US Food and Drug Administration issued a warning letter to a dietary supplement company that one of their products contained the unapproved SARM ostarine. At the time, ostarine was being investigated as a new drug candidate.[2,8]

SELECTIVE RECEPTOR MODULATION DRUGS—
HOW THEY WORK

The name of these agents accurately describes the mechanism by which they work: They modulate the activity of receptors that are naturally activated by hormones, in this case steroid structures related to testosterone (SARM) or estrogen (SERM; e.g., estradiol-beta-17) (Fig. 7.2). Steroid hormones induce their physiological effects by binding to their specific receptors and activating a cascade of biochemical events at the nuclear and cellular level—referred to as downstream events within the cells of the tissues with specific receptors (Table 7.2). Details of steroid receptor activation were presented in Chapters 2 and 4: Glucocorticoids. The unique aspect of SARMs and SERMs is that they are selective for androgenic or estrogenic receptors and are not omnifunctional across all tissues of the body.[2]

Another key aspect is related to their ability to disrupt hormonal regulator feedback loops. In endocrine function, most hormone levels in the blood are controlled by negative feedback. When a hormone level reaches a critical concentration, the hormone serves as a signaling agent to shut down its own production (Fig. 7.3 provides an illustrative example of hormonal feedback regulation). Some selective receptor modulators can block the negative feedback and allow hormone concentrations in the blood to rise to supraphysiological levels.[7]

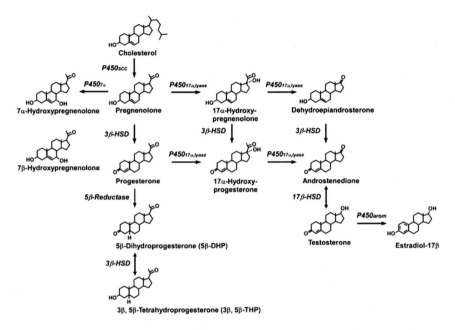

Figure 7.2 Chemical structures and biochemical production pathway of the major steroid hormones in humans (used with permission).[9]

Table 7.2 Primary Steroid Hormones in Humans and Some of their Key Physiological Functions

Hormone (Production Gland)	Target Tissues	Processes Influenced
Cortisol (Adrenal Cortex)	Muscle Adipose Liver	↑ Lipolysis, Proteolysis ↑ Lipolysis ↑ Gluconeogenesis
Testosterone (Testes, Adrenal Cortex ♂) (Adrenal Cortex ♀)	Muscle Adipose Liver Testes	↑ Protein synthesis ↑ Lipolysis ↑ Protein synthesis, Glycogenesis ↑ Spermatogenesis (♂)
Estrogen (Ovaries ♀) (Adrenal Cortex ♂)	Muscle Adipose Liver Ovaries	↑ Lipolysis, Glycogenesis ↑ Lipolysis ↑ Protein synthesis ↑ Ovulation (♀)
Progesterone (Ovaries ♀) (Adrenal Cortex ♂)	Muscle Adipose Liver Ovaries	↑ Glycogenesis ↑ Lipolysis ↑ Protein synthesis, Gluconeogenesis ↑ Ovulation (♀)

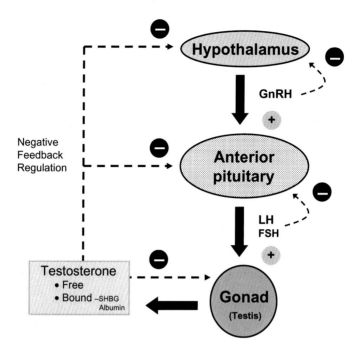

Figure 7.3 Negative feedback modulation of testosterone in males. Solid arrows indicate stimulator action (+), dashed arrows inhibitory action (−). GnRH, gonadotropin-releasing hormone; LH, luteinizing hormone; FSH, follicle-stimulating hormone; SHBG, sex hormone-binding globulin.

CONTRADICTIONS AND AMBIGUOUS EVIDENCE

SARMs

The relatively new and developing nature of SARMs as doping agents means that there is little rigorous scientific data on efficacy in sports and exercise performance, though the internet is full of testimonials on their ability to promote anabolism. In vitro and in vivo animal studies have been focused on efficacy for muscle-wasting conditions such as Duchenne muscular dystrophy and age-related sarcopenia. The compounds clearly in these cases increased muscular tissue, strength, and resistance to exercise fatigue.[10–13]

A study in women aged about 65 showed an increase in lean tissue, primarily muscle, over a 6-month period of use. The treatment group also had improved muscular strength and physical performance, but not to a greater extent than the placebo group.[14] The body composition findings were comparable to the preliminary findings of Marcantonio and associates after 3 months use (i.e., slight improvements).[15] Taken together, existing data suggest that, at the doses that have been tested, SARMs do induce modest gains in lean body mass in healthy volunteers. But the gains in skeletal muscle mass are nowhere near as great as those reported with supraphysiological doses of testosterone, which is what many doping athletes typically administer.[7] Gains of 1.0 to 1.5 kg in lean body mass with SARMs is contrasted with the 5 to 7 kg gains for 300 and 600 mg doses of testosterone enanthate, which can have serious side effects (see Chapter 2: Anabolic Androgenic Steroids).[16]

SERMs

As with SARMs, data on the effects of SERMs on skeletal muscle adaptation and exercise performance in humans are limited, though many articles show increased bone mineral density.[4] In animal studies, SERMs enhanced skeletal muscle membrane integrity, leading to a more stable cellular structure (i.e., more resistant to damage such as might occur in an intensive exercise session).[7,17] In isolated human skeletal muscle cell culture, they altered steroid receptor activation, mRNA expression, and glucose transporter 4 (GLUT4) expression.[17,18] These findings suggest that SERMs have differential effects on muscle tissue gene expression patterns and play a role in the regulation of muscle growth. How exercise may interact with these effects and relationships is unclear.

There is abundant and consistent evidence that clomiphene, tamoxifen, and raloxifene cause a reflex rise in pituitary gonadotrophin secretion and a consequent increase in blood testosterone concentration in normal men. A similar increase of 5 to 20 nmol/L was reported with aromatase inhibitors such as testolactone, exemestane, and anastrozole (see earlier discussion on aromatase inhibitor action).[16−18] An increase of this magnitude in the context of exercise training can lead to greater anabolic action with resultant increased muscle mass and strength. More research, however, is needed on this point as exact dose-response magnitude of effect can vary due to many extraneous factors such as nutrition.[7]

A key difference between male and female reproductive endocrinology is that estrogen hormones (derived in part from aromatization of testosterone) have a significant role in negative feedback regulation of blood testosterone concentration in men but not extensively in women.

The very low blood testosterone concentration in women is derived from adrenal, ovarian, and extraglandular sources and is not directly involved in substantial homeostatic feedback regulation. Hence, estrogen blockade in women would not be expected to influence adrenal or extraglandular testosterone production, and furthermore the contribution of direct ovarian testosterone secretion is small.

THROMBOSIS, EMBOLISMS, HOT FLASHES, AND HYPERANDROGENISM

SARMs

SARMs can affect anabolic and androgen processes just as testosterone and anabolic steroids do; hence, the side effects are similar although typically less severe. However, the selective action of each SARM results in varying levels of the following effects that are associated with anabolic agents:[19]

- testicular atrophy (reduced size of testicles, males)
- gynecomastia (male breast development)
- virilization (male characteristics in women)
- baldness
- acne

SERMs

SERMs may cause serious side effects, including blood clots, stroke, and possibly endometrial cancer in women. Other less severe effects are:[20]

- chest pain or shortness of breath
- fatigue
- weakness, tingling, or numbness in face, arm, or leg
- sudden difficulty seeing
- dizziness
- sudden severe headache
- leg swelling or tenderness
- night sweats or hot flashes
- mood swings
- abnormal vaginal bleeding or discharge (women)
- pain or pressure in the pelvis (women)

CONCLUSION

SARMs and SERMs are becoming popular alternatives to anabolic steroids for enhancing muscularity. They are believed by athletes to have less severe side effects. Animal studies support the assumption of positive physiological adaptation and performance enhancement, but human findings are less conclusive. These drugs are banned by WADA and national antidoping agencies. Contrary to popular belief, they do have some detrimental side effects, but not to the extent seen with more potent anabolic agents such as testosterone and anabolic steroids.

Close-Up: Selective Estrogen Receptor Modulators and Exercise as Treatments for Breast Cancer Survivors

Elizabeth S. Evans
Elon University, Elon, NC, United States

Medically, SERMs are used to treat estrogen receptor (ER)-positive breast tumors, which account for 70% of all breast cancers.[21] ERs belong to the steroid/nuclear receptor superfamily, acting as ligand-dependent transcription factors.[22] Estrogen may bind to ERs in the nucleus of the target cell in order to modulate the target cell's gene expression, or it may bind to ERs on the surface or in the cytoplasm of the target cell, thus activating second messenger systems.[21,22] The resulting cascade of cellular events produces a variety of physiological responses that depend on ER type and the specific target tissue, including those of the central nervous system, cardiovascular system, reproductive system, bone, liver, and breast.[22] ERs also prompt genes which regulate cell proliferation, and the ER-α subtype is particularly associated with the development of breast tumors.[22] SERMs compete with estrogen and have a high affinity for ERs, acting as estrogen agonists in tissues such as the liver, bone, and cardiovascular system, while acting as estrogen antagonists in the breast and uterus.[21,22]

Tamoxifen is the oldest SERM in existence and is the first-line endocrine agent for the treatment of ER-positive breast cancer in pre- and perimenopausal women, having been shown to enhance disease-free survival and reduce recurrence by 50% 15 years postdiagnosis.[22,23] Tamoxifen side effects can range in severity and may include (but are not limited to) menopause-like symptoms (e.g., hot flashes, vaginal discharge), increased risk of thromboembolic events, decreases in certain aspects of neurocognitive function, negative changes in body composition, and decreased bone mineral density.[23-27] Since a large percentage of breast cancer survivors do remain disease free a decade or more postdiagnosis, important clinical research questions exist regarding how to most accurately identify those patients who could be spared extended endocrine therapy.[25]

As an exercise physiologist who works with breast cancer survivors, I believe in using exercise interventions to mitigate the side effects of cancer treatments, increase functional capacity and quality of life, and decrease the risk of future recurrence. Since SERMs are administered for many years postdiagnosis, they have the propensity for a continual and long-lasting impact on a patient's physiology. In the case of bone health, for example, combined aerobic and resistance exercise training may help to mitigate the decreases in bone mineral density that can occur with tamoxifen use (particularly in premenopausal women), thus improving physical functioning and decreasing fall and fracture risk.[23,28] However, more research is needed to understand if the same exercise prescriptions that improve bone health in healthy populations are sufficient for improving bone health in breast cancer survivors taking tamoxifen.[28] Given the complexity of different breast cancer treatment regimes, lifestyle, physical, and other medical history factors, an individualized approach is warranted when devising exercise prescriptions that aim to improve function and quality of life in this population, including those patients using long-term SERMs.

REFERENCES

1. Smith CL, O'Malley BW. Coregulator function: a key to understanding tissue specificity of selective receptor modulators. *Endocr Rev.* 2004;25(1):45−71.

2. United States Anti-Doping Agency. Selective androgen receptor modulators (SARMS) — a prohibited class of anabolic agents. <https://www.usada.org/selective-androgen-receptor-modulators-sarms-prohibited-class-anabolic-agencts/>; 2015. Accessed 17.05.01.

3. Yin D, Gao W, Kearbey JD, et al. Pharmacodynamics of selective androgen receptor modulators. *J Pharmacol Exp Ther.* 2003;304(3):1334−1340.

4. Riggs BL, Hartmann LC. Selective estrogen-receptor modulators—mechanisms of action and application to clinical practice. *N Engl J Med.* 2003;348(7):618−629.

5. Stephenson RP. A modification of receptor theory. *Br J Pharmacol Chemother.* 1956;11 (4):379−393.

6. Geyer H, Schänzer W, Thevis M. Anabolic agents: recent strategies for their detection and protection from inadvertent doping. *Br J Sports Med.* 2014;48:820−826.

7. Hackney AC, Anderson T, Dobridge J. Exercise and male hypogonadism: testosterone, the hypothalamic-pituitary-testicular axis, and exercise training. In: Winters S, Huhtaniemi I, eds. *Male Hypogonadism: Basic, Clinical and Therapeutic Principles.* New York: Springer; 2017:257−280.

8. U.S. Food and Drug Administration, Department of Health and Human Services. Biogenix USA, LLC Warning Letter. December 11, 2014. <http://www.fda.gov/ICECI/ EnforcementActions/WarningLetters/ucm434928.htm>; 2014. Accessed 17.05.23.

9. Tsutsui K. Neurosteroid biosynthesis and function in the brain of domestic birds. *Front Endocrinol.* 2011;2:37.

10. Cozzoli A, Capogrosso RF, Sblendorio VT, et al. GLPG0492, a novel selective androgen receptor modulator, improves muscle performance in the exercised-mdx mouse model of muscular dystrophy. *Pharmacol Res.* 2013;72:9−24.

11. Akita K, Harada K, Ichihara J, et al. A novel selective androgen receptor modulator, NEP28, is efficacious in muscle and brain without serious side effects on prostate. *Eur J Pharmacol.* 2013;720(1−3):107−114.

12. Gao W, Reiser PJ, Coss CC, et al. Selective androgen receptor modulator treatment improves muscle strength and body composition and prevents bone loss in orchidectomized rats. *Endocrinology.* 2005;146:4887−4897.

13. Miner JN, Chang W, Chapman MS, et al. An orally active selective androgen receptor modulator is efficacious on bone, muscle, and sex function with reduced impact on prostate. *Endocrinology.* 2007;148(1):363−373.

14. Papanicolaou DA, Ather SN, Zhu H, et al. A phase IIA randomized, placebo-controlled clinical trial to study the efficacy and safety of the selective androgen receptor modulator (SARM), MK-0773 in female participants with sarcopenia. *J Nutr Health Aging.* 2013;17 (6):533−543.

15. Marcantonio EE, Witter RE, Ding Y, et al. A 12-week pharmacokinetic and pharmacodynamic study of two selective androgen receptor modulators (SARMs) in postmenopausal subjects. *Endocr Rev Suppl.* 2010;31(1):872.

16. Bhasin S, Jasuja R. Selective androgen receptor modulators as function promoting therapies. *Curr Opin Clin Nutr Metab Care.* 2009;12(3):232−240.

17. Handelsman DJ. Indirect androgen doping by oestrogen blockade in sports. *Br J Pharmacol.* 2008;154(3):598−605.

18. Handelsman DJ. The rationale for banning human chorionic gonadotropin and estrogen blockers in sport. *J Clin Endocrinol Metab.* 2006;91(5):1646−1653.

19. U.S. Department of Defense, Human Performance Resource Center. What are SARMs and are They Safe to Use as Dietary Supplements? <http://hprc-online.org/dietary-supplements/ opss/operation-supplement-safety-OPSS/opss-frequently-asked-questions-faqs-1/what-are-sarms-and-are-they-safe-to-use-as-dietary-supplements>; 2015. Accessed 17.05.01.

20. Selective Estrogen Receptor Modulators (SERMs). <http://www.breastcancer.org/treatment/ hormonal/serms>; 2017. Accessed 17.07.01.

21. Jeselsohn R, De Angelis C, Brown M, et al. The evolving role of the estrogen receptor mutations in endocrine therapy-resistant breast cancer. *Curr Oncol Rep.* 2017;19:35.

22. Jameera Begam A, Jubie S, Nanjan MJ. Estrogen receptor agonists/antagonists in breast cancer therapy: a critical review. *Bioorg Chem.* 2017;71:257−274.

23. Zeidan B, Anderson K, Peiris L, et al. The impact of tamoxifen brand switch on side effects and patient compliance in hormone receptor positive breast cancer patients. *Breast.* 2016;29:62−67.

24. Gu R, Weijuan J, Zeng Y, et al. A comparison of survival outcomes and side effects of toremifene or tamoxifen therapy in premenopausal estrogen and progesterone receptor positive breast cancer patients: a retrospective cohort study. *BMC Cancer.* 2012;12:161.

25. Ribnikar D, Sousa B, Cufer T, et al. Extended adjuvant endocrine therapy − a standard to all or some? *Breast.* 2017;32:112−118.

26. Schilder CM, Seynaeve C, Beex LV, et al. Effects of tamoxifen and exemestane on cognitive functioning of postmenopausal patients with breast cancer: results from the neuropsychological side study of the tamoxifen and exemestane adjuvant multinational trial. *J Clin Oncol.* 2010;28:1294–1300.

27. Hojan K, Milecki P, Molinska-Glura M, et al. Effect of physical activity on bone strength and body composition in breast cancer premenopausal women during endocrine therapy. *Eur J Phys Rehabil Med.* 2013;49:331–339.

28. Knobf MT, Jeon S, Smith B, et al. Effect of a randomized controlled exercise trial on bone outcomes: influence of adjuvant endocrine therapy. *Breast Cancer Res Treat.* 2016;155:491–500.

CHAPTER 8

Narcotics

Narcotics are drugs that act pharmacologically like morphine, a constituent of the opium poppy (*Papaver somniferum*), meaning that they change the psychic and physical status of a person to reduce pain (analgesia), induce sleep, and alter mood or behavior (e.g., induce euphoria). The category includes all natural, semisynthetic, and synthetic opioids that act pharmacologically like morphine.[1] The various types of opioids are all on the Prohibited List of the World Anti-Doping Agency (WADA). In some jurisdictions, "narcotic" also refers to cocaine, which chemically is not a narcotic but a sympathomimetic (see Chapter 3) having catecholamine (epinephrine and norepinephrine) actions; WADA classifies them as stimulants. Drugs that elicit psychic changes like those manifested in psychosis—psychotomimetics, psychedelics, hallucinogens—also are sometimes mistakenly called narcotics; of these, only cannabinoids are banned by WADA. In this chapter, "narcotics" refers only to opioids.[2]

IN THE ARMS OF MORPHEUS

The opioids have a long history. Ancient Egyptian hieroglyphics tell of the poppy being used for pain relief during childbirth. Around 400 BCE, Hippocrates, the Greek father of medicine, wrote of the varied medical uses of extracts from poppies. In the early 16th century, Paracelsus made an opium pill (it also contained citrus juice and gold) and a tincture (alcohol extract) of opium called laudanum (from the Latin *laudare*, to praise), which is still available in some countries.[3,4] In 1817, the German chemist Friedrich Wilhelm Adam Serturner isolated the active ingredient from opium and called it morphine, after Morpheus, the Greek god of dreams. In the 1850s, with the development of the hypodermic needle and syringe, opiates could be administered intravenously, increasing their efficacy and potency and their use for the treatment of pain. In the 20th century, synthetic and semisynthetic forms of morphine were developed. Pethidine (meperidine), one of the first pure synthetics, was produced in Germany in 1939. It was

Doping, Performance-Enhancing Drugs, and Hormones in Sport. DOI: https://doi.org/10.1016/B978-0-12-813442-9.00008-0

widely used in the mid-20th century until the development of agents with greater potency and milder side effects.[3-5] In the 19th and early 20th centuries, opium, morphine, and related compounds were ingredients of many pharmaceutical products, sometimes unbeknownst to the consumer. For example, Mrs. Winslow's Soothing Syrup for teething and colicky babies, unlabeled yet containing morphine, killed many infants.[6] Public outcries about such tragedies ultimately resulted in the 1938 Federal Food, Drug, and Cosmetic Act in the United States. It required that medications to be proven safe by the Food and Drug Administration, which included information to consumers on content. The Controlled Substances Act of 1970 imposed greater regulation and scheduling of drugs with abuse potential. As noted in Chapter 3, the US Drug Enforcement Agency (DEA) classifies drugs into five schedules according to their potential for abuse, medical applications, and safety. Schedule I drugs have the highest abuse potential and (according to the law) no current medical use. Heroin is in schedule I; morphine, oxycodone, and hydrocodone are in schedule II (i.e., opioid-type drugs).[7]

In the first decade of the 21st century, healthcare policymakers increased the emphasis on controlling and managing pain. This approach led to a large increase in opioid prescriptions. Some newly developed agents such as Percodan, Percocet, OxyContin, Lortab, and Vicodin were overprescribed, leading to a rise in addiction and hundreds of thousands of emergency department visits due to medication abuse. The pharmaceutical industry responded with abuse-deterrent formulations, but that does not prevent oral abuse, which in some places is currently epidemic and socially devastating.[4,7]

THE ATHLETIC NEED TO PUSH THROUGH THE PAIN

Exercise training and competition in sport are physically demanding, and many athletes live with some degree of acute or chronic pain.[8] High doses of narcotics to relieve the pain during training programs could lead to enhanced performance, as could masking pain during a competitive event. In this context, narcotics, by altering the pain threshold, are potential performance-enhancing and enabling drugs (Fig. 8.1), which is why they are banned by WADA except where a therapeutic use exemption is granted for medical reasons or medical standard of care dictates the usage (see the close-up at the end of this chapter).[2]

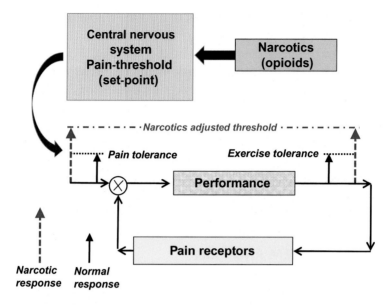

Figure 8.1 How narcotics alter the central nervous system pain threshold by interacting with peripheral pain receptors (nociceptors). The ⊗ (operator) symbol denotes an action occurrence.

Use of narcotics is rampant in physically demanding and combative-type sports. Cottler and associates surveyed retired National Football League players and found that over 50% had regularly used opioids during their professional careers to deal with chronic pain. Sixty-three percent of the users obtained them from a nonmedical source—teammate, coach, trainer, family member, dealer, or the Internet.[9] Young nonprofessional athletes in a variety of sports have similar experiences with pain and also use narcotics for it.[10]

This trend is of course not limited to athletes. Prescriptions of opioid pain relievers skyrocketed in the United States over the past 25 years. Nor is use limited to adult nonathletes, as nearly 1 in 4 high school seniors has had some exposure to prescription opioids either medically or nonmedically.[11] The number of narcotic prescriptions rose from 76 million in 1991 to 207 million in 2013, with the United States being the biggest consumer globally, accounting for almost 100% of the world's total for hydrocodone and 81% for oxycodone.[12] Furthermore, drug deaths in the United States appear to be rising faster than ever ($\sim$60,000 in 2016) due in large part to the opioid fentanyl, which is about 100 times more potent than heroin.[13]

There have been high-profile cases of athletes whose abuse of narcotics ended in tragedy. Tom McHale, who played for the Tampa Bay Buccaneers and Philadelphia Eagles in the National Football League, died in 2008 of an accidental overdose. Craig Newsome, a former Green Bay Packers defensive back, told the *Milwaukee Journal Sentinel* in 2012 that he became severely addict during his career and postcareer and was willing to do whatever was necessary to continue getting pain pills. Derek Boogaard, a National Hockey League "enforcer," became addicted, and a mix of oxycodone and alcohol led to his death in 2011. A 2011 study noted that narcotic abuse and addiction rates by football players were three times that for men of similar age.[9,12,14]

THE OPIOID SYSTEM AND PAIN

Opioids act by binding to opioid receptors on nerve cells (neurons) in the central (brain and spinal cord) and peripheral nervous systems and reducing the neuron excitability that leads to the sensation of pain. There are three main opioid receptor types, each with select subtypes:[15,16]

- mu (μ_1, μ_2 μ_3)
- delta (δ_1, δ_2)
- kappa (κ_1, κ_2, κ_3)

Activation of the specific receptors produces a variety of distinct psychophysiological effects (Table 8.1), the most critical to sport being analgesia. The mechanism by which opioids reduce pain involves their intracellular actions on neurons. When an opioid binds to its receptor in the neuronal membrane, calcium channels close, blocking positively charged calcium ions from entering the neuron cell. As a result, the level of the intracellular signaling agent cyclic adenosine monophosphate (cAMP) decreases, and potassium channels open, allowing positive potassium ions to exit the cell. These events hyperpolarize the cell, increasing the electrical charge difference between the cell's interior and the extracellular environment and making the neuron less likely to send the neuronal signal called an action potential. This change in electrical polarity prevents the action potential pain signal from moving forward in the neural circuit, quieting the neurons along the pain pathway; that is, the opioid dampens the transmission of a pain signal to the brain.[15] Peripheral analgesic action is also partly related to

Table 8.1 Opioid Receptor Subtypes Associated With Central and Peripheral Nervous System Effects

Receptor Type	Activation Effect
Delta (δ)	Analgesia Antidepressant Convulsions Physical dependence μ-Mediated respiratory depression (?)
Kappa (κ)	Analgesia Anticonvulsant Depression Dissociative/hallucinogenic Diuresis Dysphoria Miosis (pupillary constriction) Neuroprotection Sedation Stress (anxiety)
Mu (μ)	Analgesia Physical dependence Respiratory depression Miosis Euphoria Reduced GI motility Physical dependence Vasodilation (?)
(?) = unclear whether the effect is induced directly by receptor activation or indirectly via other mechanisms.	

histamine release. Histamine stimulates the nociceptors (pain receptors), but its effect is experienced as an itch rather than a pain.[15–17]

Euphoria results from stimulation of the brain's reward system involving the neuronal mu receptors and dopamine. Dopamine is a neurotransmitter present in regions of the brain that regulate movement, emotion, cognition, motivation, and pleasure. The overstimulation of this neural system, which rewards natural positive behaviors, produces euphoria. Activation of mu receptors in the brain results in hyperpolarization at the neuron (see earlier discussion), but this electrical change results in reduced γ-aminobutyric acid (GABA) release from the neurons. GABA ordinarily inhibits the release of dopamine; hence, opioids allow greater dopamine release and the resulting sense of elation and pleasure.[15–17]

Other analgesics associated with sport are narcotic-like. Tramadol, for example, works by affecting how the brain responds to or perceives pain, causing an increase in the neurotransmitters norepinephrine and serotonin and endorphins. Endorphins are endogenous substances—

the body's own opiates, as it were—that alleviate the sense of pain. Tramadol works by slightly different yet overlapping mechanisms as true opioids, causing some pharmacologist to suggest it should be classified as a narcotic.[18] But even though it has opiate receptor ligand activity and is addictive, it is only a DEA Schedule IV drug. Narcotics are banned by WADA [S7 category of performance-enhancing drugs (PEDs)], but tramadol is not. Several national antidoping agencies (the United States Anti-Doping Agency is one) believe that the time has come for WADA to move this drug from its Monitoring Program, where it has been since 2012, to the Prohibited List.

SUBSTANCES THAT WORK TOO WELL

There is a plethora of research on the psychological and physiological implications of narcotic use dating back over a hundred years, with one of the first scientific mentions being in 1889 in the journal *Science*. Most of the scientific literature concerning athletes and nonathletes since then has focused on the addictive rather than analgesia properties, perhaps because the analgesic effect has been taken for granted for centuries. Morphine was a mainstay in the Western pharmacopoeia in the 19th century for treatment of pain, anxiety, respiratory problems, "consumption," "women's ailments," and medical problems in general.[6]

The American Civil War marked a turning point in attitude towards narcotics. So many soldiers became addicts that morphine addiction was known as the Soldier's disease. Heroin was synthesized in 1898, and the manufacturer, Bayer, offered it as a cough suppressant and nonaddictive morphine substitute. The side effects of morphine and heroin, especially addiction, had become so apparent by the turn of the century that public action was demanded.[4,7] In 1914, the Harrison Narcotics Tax Act made distribution by doctors and pharmacists somewhat more difficult. Bayer ultimately stopped mass production of heroin, and the Anti-Heroin Act of 1924 made the drug illegal in the United States even for medical use.[6]

In an attempt to maintain the painkilling effects of morphine and heroin but lessen the risk of dependence, German pharmacologists produced the semisynthetic oxycodone in 1916, although it did not achieve global distribution until the late 1930s. Shortly after World War II, numerous "nerve block clinics" were opened using this drug to treat the pain of injured soldiers.[7] But it too proved addictive and problematic.

Also after World War II, scientific research was done on the use of natural and synthetic opioids to reduce the pain of specific medical procedures. The results were highly positive, but many of the published reports glossed over the side effects.[19,20] In the 1950s, research shifted to understanding the mechanism of action.[21] Some medical professionals remained hesitant about prescribing opioids because of the side effects, but the World Health Organization (WHO) endorsed the medical use of morphine in 1969, saying that it "doesn't inevitably lead to dependence." This statement came about because the WHO distinguished between physical dependence and "drug dependence," which they defined as "difficulty controlling consumption, compulsive use, and inappropriate social behaviors."[7,22] This line of thought was supported by clinicians and scientists. In 1980, Dr. Hershel Jick published a letter in the *New England Journal of Medicine* stating that "the development of addiction is rare in medical patients with no history of addiction."[23] Shortly thereafter, Drs. Russell Portenoy and Kathleen Foley reported in the journal *Pain* that "opioid maintenance therapy can be a safe, salutary and more humane alternative to the options of surgery or no treatment in those patients with intractable non-malignant pain and no history of drug abuse."[24] Comments such as these led many health professionals to push toward a priority of humane pain treatment by the prescriptive use of opioids and their derivatives. In the 1990s, pharmaceutical companies responded by developing extended-release formulations, a primary example being OxyContin.

By the start of the 21st century, opioids were overprescribed to the point where prescription analgesic poisoning was a more common cause of death than heroin or cocaine overdose. By 2010, overdose deaths due to prescription opioids had more than tripled in 20 years, causing 16,651 deaths in the United States. Many early opioid advocates now admit that they had been wrong. In 2011, Dr. Portenoy stated, "Clearly if I had an inkling of what I know now then, I wouldn't have spoken in the way that I spoke. It was clearly the wrong thing to do."[7]

ADDICTION, ADDICTION, AND ADDICTION

The side effects of narcotics are copiously documented. The most menacing are overdosing and physical dependency.[3,16,25] Table 8.2 lists the more and less common acute effects. Not shown are the psychophysiological effects that can have a devastating impact on the social and economic welfare of the individual. The United States National

Table 8.2 Acute Opioid-Induced Side Effects	
More Common Side Effects	**Less Common Side Effects**
Constipation	Urinary retention
Nausea	Pruritis (itching)
Sedation	Delirium
Confusion	Myoclonus (muscle twitches)
Hallucinations	Hyperalgesia (heightened pain sensitivity)
Sweats	Seizures
Dry mouth	Respiratory depression

Institute on Drug Abuse (NIDA) statistics for 2013 estimated the costs related to crime, lost work productivity, and health care at more than US$78 billion.[12] Testifying before the Senate Caucus on International Narcotics Control, NIDA Director Dr. Nora Volkow stated that the abuse of and addiction to opioids is a serious global problem affecting all societies, with an estimated 26 to 36 million abusers. For the United States, an estimated 2.1 million people had a substance use disorders related to opioid pharmaceuticals in 2012, as contrasted with 467,000 addicted to heroin.[12]

CONCLUSION

Exercise training is physically demanding and results in many athletes experiencing persistent pain and discomfort. The analgesic effect of opioids is a powerful incentive for their use. They are classified as PEDs and banned by antidoping agencies such as WADA. They do have legitimate therapeutic medical uses in athletes, primarily in the treatment of injuries, but physicians must follow strict guidelines of use and reporting. They are highly addictive, and athletes using them as PEDs or for legitimate medical purposes are not immune to the negative side effects, including addiction.

Close-Up: Pain Management in Athletes—How do Physicians Deal with this Issue when Narcotics are WADA Banned Substances?

By Maarit Valtonen
Medical Director, Finnish Olympic Committee, Jyväskylä, Finland

This is a personal story. "Please bring me some efficient pain killers so I can sleep" states the text message. I am immediately concerned. Her ruptured lumbar disc had become symptomatic during the competitive season. Rigorous physiotherapy had been, to-date, sufficient to continue training and keep symptoms in balance. But the World Championships will be soon upon us, and all focus is directed to the most important competition of the season. As long as there are no signs of motor paresthesia, the athlete wants to go all out and compete. The text, however, tells me something has changed.

Maybe traveling to the preparation camp has irritated the nerve again? No other symptoms, just the pain, the athlete reports, and the training sessions had to-date been great. I have a feeling that everything is not as it should be. I grab some strong painkillers on the way to the airport, just in case. I am very familiar with the WADA regulations and ethics concerning the use of these substances. However, as a team physician, my first priority is to treat my patient with correct medical ethics and give them optimal medical care.

When I arrive at the camp, the situation is not looking great. Symptoms have gone from bad to worse over the past 24 hours. No motor impairment, but there is severe radicular pain in the buttock, leg, and thigh. The athlete cannot get out of the hotel room. I start to organize a transfer to a local hospital. Through the help of our international network, we find the best medical facility for the athlete. However, it is a weekend, and we need to handle the situation in the hotel until Monday morning.

I explain, to the athlete the risk of using these medications if testing authorities arrive. There is no way they can even get a sample in my current situation, is the screamed response. A clear indication of the severity of the situation! I consult the Finnish Anti-Doping Agency guidelines and go through the correct protocol. There is no need to complete a therapeutic use exemption application at this point: This is a medical emergency. I carefully document each and every dose of medication and keep track of the daily limits.

After two challenging days at the "field hospital," I get my athlete to the neurosurgeon recommended to us. I feel a bit relieved as the athlete is admitted to in-patient care. From now on, the hospital rules apply. Nevertheless, the athlete is still obligated to report her whereabouts to the WADA authorities and keep a list of every substance administered. I remind the athlete of this requirement.

Sports team physicians are beyond neurotic to follow the WADA regulations. We do not want to harm our athlete's career through our medical treatment. However, every athlete is entitled to the best medical care they require and need. Rigorous reporting, being familiar with the regulations, and consulting the antidoping agency assist physicians to offer the best medical care within the WADA regulations. But it is an ongoing and demanding challenge for the sports physician in practicing medicine.

REFERENCES

1. Mangione MP, Matoka M. Improving pain management communication: how patients understand the terms "opioid" and "narcotic." *J Gen Intern Med*. 2008;23(9):1336−1338.

2. The World Anti-Doping Code International Standard − Prohibited List 2017. 2017. ⟨https://www.wada-ama.org/sites/default/files/resources/files/2016-09-29_-_wada_prohibited_list_2017_eng_final.pdf⟩; 2017 Accessed 07.14.2017.

3. Gahlinger P. *Illegal Drugs: A Complete Guide to Their History, Chemistry, Use, and Abuse*. New York: Plume Publications; 2004.

4. Meldrum ML. A capsule history of pain management. *JAMA*. 2003;290(18):240−245.

5. Northwestern University School of Professional Studies. Exploring topics in sports: why do athletes risk using performance enhancing drugs? ⟨http://sps.northwestern.edu/main/news-stories/why-do-athletes-risk-using-peds.php⟩; 2015 Accessed 07.15.2017.

6. Food and Drug Administration: Significant Dates in U.S. Food and Drug Law History. ⟨https://www.fda.gov/aboutfda/whatwedo/history/milestones/ucm128305.htm⟩; 2014 Accessed 06.28.2017.

7. Sheikh N. The History of Opioids in America: Pain Patients or Prescription Addictions? ⟨https://sobernation.com/the-history-of-opioids-in-america-pain-patients-or-prescription-addictions/⟩; 2016 Accessed 07.01.2017.

8. Hackney AC. *Exercise, Sport and Bioanalytical Chemistry: Principles and Practice*. New York, New York: Elsevier − RTI Press; 2016.

9. Cottler LB, Abdallah AB, Cummings SM, et al. Injury, pain, and prescription opioid use among former National Football League (NFL) players. *Drug Alcohol Depend*. 2011;116(1−3):188−194.

10. Stache S, Close JD, Mehallo C, et al. Nonprescription pain medication use in collegiate athletes: a comparison of samples. *Physician Sportsmed*. 2014;42(2):19−26.

11. McCabe SE, West BT, Teter CJ, et al. Medical and nonmedical use of prescription opioids among high school seniors in the United States. *Arch Pediatr Adolesc Med*. 2012;166(9):797−802.

12. National Institute of Drug Abuse. America's Addiction to Opioids: Heroin and Prescription Drug Abuse. ⟨https://www.drugabuse.gov/about-nida/legislative-activities/testimony-to-congress/2016/americas-addiction-to-opioids-heroin-prescription-drug-abuse⟩; 2014 Accessed 05.01.2017.

13. Katz J. NYTimes − Drug Deaths in America Are Rising Faster Than Ever. June 5, 2017. ⟨https://www.nytimes.com/interactive/2017/06/05/upshot/opioid-epidemic-drug-overdose-deaths-are-rising-faster-than-ever.html⟩; 2017 Accessed 07.03.2017.

14. Wertheim LJ, Rodriguez K. Smack epidemic. *Sports Illustrated*. June 22, 2015. ⟨https://www.si.com/more-sports/2015/06/18/special-report-painkillers-young-athletes-heroin-addicts⟩; 2015 Accessed 07.01.2017.

15. Dhawan BN, Cesselin F, Raghubir R, et al. International Union of Pharmacology XII: classification of opioid receptors. *Pharmacol Rev*. 1996;48(4):567−592.

16. Drugs.com. Understanding Opioid (Narcotic) Pain Medications. ⟨https://www.drugs.com/article/opioid-narcotics.html⟩; 2016 Accessed 03.30.2017.

17. Harrison S, Geppetti P, Substance P. *Int J Biochem Cell Biol.* 2001;33(6):555—576.

18. World Health Organization — Tramadol: Update Review Report. Expert Committee on Drug Dependence, 36th Meeting, Geneva, 2014.

19. Maximov A. Demerol analgesia in obstetrics. *Calif Med.* 1946;65(2):43—47.

20. Monroe MW. Medical pain relief; clinical observations of a morphine derivative, metapon hydrochloride. *Marquette Med Rev.* 1949;14(2):53.

21. De Jongh DK. Remarks on the mechanism of analgesic action of morphine. *Acta Physiol Pharmacol Neerl.* 1954;3(2):164—172.

22. World Health Organization — WHO Expert Committee on Drug Dependence. Technical Report Series, No. 407, Geneva, 1969.

23. Jacobs H. This One-Paragraph Letter May Have Launched the Opioid Epidemic. ⟨http://www.businessinsider.com/porter-and-jick-letter-launched-the-opioid-epidemic-2016-5⟩; 2016 Accessed 07.10.2017.

24. Portenoy RK, Foley KM. Chronic use of opioid analgesics in non-malignant pain: report of 38 cases. *Pain.* 1986;25(2):171—186.

25. Rosenblum A, Marsch LA, Joseph H, et al. Opioids and the treatment of chronic pain: controversies, current status, and future directions. *Exp Clin Psychopharmacol.* 2008;16(5):405—416.

CHAPTER 9

Beta Blockers

Beta blockers—beta-adrenergic receptor blockers—are used primarily to treat heart disease and related conditions. These drugs reduce blood pressure and manage cardiac arrhythmias and are cardioprotective after myocardial infarction (heart attack).[1] Beta blockers bind to beta-adrenoceptors on cells and thereby block the binding of norepinephrine and epinephrine. Norepinephrine is the major neurotransmitter of the sympathetic nervous system. It is released into the periphery upon activation of the sympathetic system, which is the major neural regulator of cardiovascular function. Epinephrine is released from the adrenal medullary gland and also has influence on the heart and blood vessels. These two substances, known as catecholamines, have roles not only as neurotransmitters but as hormones due to their endocrine actions.[2] The binding of beta blockers to adrenoreceptors leads to a reduction in heart rate and strength of the heart muscle contraction. This produces a general sense of relaxation and calmness,[1,3,4] which is advantageous in sports where nervousness and heightened anxiety interfere with performance—e.g., golf, archery, and shooting. Therefore, beta blockers are performance-enhancing drugs (PEDs) and are banned by the World Anti-Doping Agency (WADA).

CARDIOVASCULAR MODULATION AND HEALTH

Beta blockers are competitive pharmacologic inhibitors of catecholamines. More than a century ago, investigators deduced that catecholamines were neurotransmitter-hormonal substances, excitatory in nature, that acted by binding selectively to receptor-like structures on tissues to produce their physiological actions. This revelation came from the initial catecholamine research begun in the late 19th century in Great Britain by Dr. George Oliver and Sir Edward Albert Sharpey-Schafer (the founder of endocrinology) in their work with a pharmacologically active extract taken from the adrenal glands.[1] In 1948, Dr. Raymond Ahlquist in the United States conducted a series

Doping, Performance-Enhancing Drugs, and Hormones in Sport. DOI: https://doi.org/10.1016/B978-0-12-813442-9.00009-2

of classic experiments which determined that there were two distinct tissue-organ responses to catecholamine-like drugs: alpha- and beta-adrenergic receptor mediated actions.[5] In 1958, dichloroisoproterenol, the first beta-blocker drug, was produced by the pharmaceutical company Eli Lilly. While somewhat effective, it was relatively nonspecific and had low potency. In the early 1960s, Sir James Black (who received the 1988 Nobel Prize in Physiology or Medicine) and associates at Imperial Chemical Industries in Great Britain developed the beta-adrenergic blocker propranolol, a major drug still in use. It reduced myocardial oxygen consumption (a critical physiological action to treat angina pectoris), hypertension, and cardiac arrhythmia. This was the first major advance in the treatment of angina since the introduction of nitroglycerin almost 100 years earlier.[1,4,6]

Research in the 1970s and 1980s led to the introduction of drugs with relative selectivity for cardiac beta-1 receptors (metoprolol, atenolol), partial adrenergic agonist activity (pindolol), concomitant alpha-adrenergic blocking activity (labetalol, carvedilol), and additional direct vasodilator activity (nebivolol). Long-acting and ultra-short-acting formulations of beta blockers were developed.[4] In the development process, pharmacology has come a long way, as the first-generation drugs were relatively nonselective, meaning that they blocked both beta-1 and beta-2. Second-generation drugs were more cardioselective, targeting primarily beta-1, though selectivity is lost at higher doses. Third-generation drugs have a vasodilator action through blockade of vascular alpha receptors, thereby reducing blood pressure (Table 9.1).[8]

Applications of beta blockers now extend beyond the cardiovascular system to prevention of migraine headache, treatment of benign essential tremor for patients with pheochromocytoma and thyrotoxicosis, and topical formulations to reduce intraocular pressure in open-angle glaucoma. They are also used to reduce portal hypertension in patients with liver cirrhosis and to reduce delirium tremens and stage fright (see the following section). Beta-1 receptors are present in the kidneys, where they control the release of the hormone renin, which increases blood pressure, so beta-1 blockade of kidney receptors reduces blood pressure.[4]

Over decades of clinical use, beta blockers have demonstrated good safety in patients of all ages, and they can be prescribed in combination with other drug classes without causing severe interaction effects.

Table 9.1 Some Commonly Prescribed Beta Blockers (generic name in brackets)
Sectral (acebutolol)
Tenormin (atenolol)
Kerlone (betaxolol)
Zebeta and Ziac (bisoprolol)
Coreg (carvedilol)
Normodyne and Trandate (labetalol)
Lopressor and Toprol-XL (metoprolol)
Corgard (nadolol)
Bystolic (nebivolol)
Levatol (penbutolol)
Visken (pindolol)
Inderal and Inderal LA (propranolol)
Blocadren (timolol)
Brevibloc (esmolol)
Beta blockers are the fourth most commonly prescribed class of drugs in the United States, with nearly 200 million prescriptions annually.[7]

STAYING CALM, COOL, AND COLLECTED ON THE BATTLEFIELD OF SPORT

Beta blockers were designed primarily to improve heart function under conditions of an impaired cardiovascular system, but they also have a powerful calming effect and stabilize motor performance, which is beneficial in sports requiring precision such as golf, bowling, darts, billiards, shooting, and archery.[9] In 1985, the International Olympic Committee put beta blockers on its Prohibited List. Initially, they were still permitted for medical treatment, but they were completely banned starting with 1988. They are prohibited from use only in competition for most sports, but for shooting and archery they are prohibited outside of competition as well. Details can be found on the WADA List.[9]

As with other PEDs, some athletes have defied the ban and suffered the consequences. Doug Barron, an American journeyman professional golfer, was the first to be banned under the PGA Tour's antidoping policy after testing positive for beta blockers and testosterone. Bill Werbeniuk, a Canadian professional snooker and pool player, was banned for using propranolol. Perhaps, the most famous case was North Korean sport shooter Kim Jong-su, who tested positive for propranolol 3 days after winning the silver and bronze in pistol shooting at the 2008 Beijing Olympics (see the close-up at the end of this chapter).[10]

BLOCKING THE ACTIONS OF NEUROTRANSMITTERS AND HORMONES

Beta blockers bind to beta receptors located in aspects of the heart (cardiac nodal tissue, conducting system, and contracting myocytes [heart muscle cells]).[3] The heart has both beta-1 and beta-2 receptors, with beta-1 predominant in number and function.[11] Beta-1 receptors primarily bind norepinephrine that is released from sympathetic adrenergic neurons. They also bind norepinephrine and epinephrine circulating in the blood, released from the adrenal medullary gland or as sympathetic neural 'spillover'.[11] Beta blockers prevent the norepinephrine or epinephrine from binding by competing for the binding site (antagonist action, see Chapter 7: Hormone and Metabolic Modulators).

Biochemically, beta receptors are coupled to G proteins that activate adenylyl cyclase to form cyclic adenosine monophosphate (cAMP) from adenosine triphosphate. Increased cAMP activates a cAMP-dependent protein kinase (PKA) that phosphorylates L-type calcium channels, which causes increased calcium entry into the cell. Increased calcium entry during a neuronal action potential leads to enhanced release of calcium by the sarcoplasmic reticulum organelles in the heart myocytes. These combined actions increase contractility (strength of contraction) and heart rate (number of contractions), and these actions in turn increased the energy and oxygen requirements of the tissue (i.e., increase the work of the heart). The activated PKA also phosphorylates sites on the sarcoplasmic reticulum, which leads to the enhanced release of calcium therefrom. This provides more calcium for binding to the contractile proteins of the myocytes such as troponin, which leads to enhanced muscular contractility. Beta blockers slow or lessen all of these actions and mitigate the work of the heart at the cellular level.[6]

Vascular smooth muscle has principally beta-2 receptors, although alpha receptors exist in these tissues too (see below). These beta receptors, like those in the heart, are coupled to a G protein to stimulate the formation of cAMP. In contrast to the heart, in vascular smooth muscle an increase in cAMP relaxes the muscle and reduces contractile force. Interestingly, compared to their effects in the heart, beta blockers have relatively little overall vascular effect because beta-2 receptors have only a small modulatory role on basal vascular tone.[4,6] Nevertheless, blockade of beta-2 receptors is associated with a small

Table 9.2 Effects of Beta and Alpha Blockers on Cardiovascular Parameters (Early = Initial Usage; Late = Prolonged Usage)		
Physiological Parameter	Beta Blocker (e.g., propranolol)	Alpha Blocker (e.g., labetalol)
Heart rate	Decreased	Unchanged
Cardiac output	Decreased	Unchanged or decreased
Venous tone	Unchanged	Decreased
Postural hypotension	Negligible	Evident
Renal blood flow	Decreased (early)	Unchanged
	Normal (late)	
Peripheral vascular resistance	Increased (early)	Decreased
	Decreased (late)	
Antihypertensive efficacy	Good	High

degree of vasoconstriction in many vascular beds. This occurs because beta blockers remove the small beta-2 receptor vasodilator influence that is normally opposing the more dominant alpha-receptor mediated vasoconstrictor influence. But as noted, beta-1 receptors are also present in the kidneys, where they control the release of renin, which increases blood pressure; hence, the blocking actions of the drugs can lead to a major reduction in blood pressure through this mechanism. Furthermore, third-generation beta blockers also have vasodilator actions through blockade of vascular alpha receptors, thereby further reducing blood pressure. Table 9.2 contrasts the physiological effects of beta and alpha blockade.

STRONG EVIDENCE OF PERFORMANCE ENHANCEMENT

In the mid-1970s, a team of British researchers conducted a unique study of nonclinical uses of beta blockers. They tested effects on the performance of skilled string musicians. They induced stress by booking the subjects into an imposing concert hall, invited the press, and recorded the sessions. The intent was to create a scenario that would produce high levels of stress and potential stage fright. The musicians were asked to perform four times, twice on placebo and twice on a beta blocker, and performances were scored by professional judges. Not only did the musicians experience less muscle tremor when on the beta blocker, they also played better. The general improvement was minimal, but for a few subjects it was dramatic.[12]

In a similar but more sport-specific study, Swedish researchers found that beta blockers improved the performance of a group of shooters in a noncompetitive setting by about 13%. The improvement was ascribed primarily to reduction of hand muscle tremor. What was not determined was whether the drugs helped nervous shooters more than calm ones, and whether the effect would have been different if the shooters had performed in a stress-inducing public competition.[13] Nonetheless, the magnitude of improvement was highly meaningful for this type of competition, where an improvement of less than 1% can make the difference between winning an Olympic medal and just being a participant.[14] Subsequent research supported the efficacy of beta blockers to improve sports performance in activities were a calm demeanor was essential for improvement of accuracy.[15]

With respect to the cardiovascular response to aerobic exercise, Hawkins and associates[16] showed that a beta blocker reduced cardiac output—amount of blood pumped by the heart per minute[11]—in healthy subjects during moderate and intensive aerobic exercise. Reduction of cardiac output is beneficial for people with heart disease, but not for healthy people doing aerobic exercise and certainly not for competitive athletes. A marathon runner or other endurance athlete with precompetition anxiety would be calmed by beta blockers, but performance would suffer and they obviously could be accused of doping. This is regrettable, as there are many anecdotal reports of athletes suffering from anxiety and panic severe enough to cripple their ability to compete.[17]

FROM LETHARGY TO IMPOTENCE: UNDESIRABLE EFFECTS

The most important side effects of beta blockers are related to their influence on cardiac function and mechanisms via a reduction in the effectiveness of the sympathetic nervous system:[4,5,7]

- lower heart rate (bradycardia)
- reduced exercise capacity
- heart failure (rare)
- low blood pressure (hypotension)
- orthostatic intolerance (lightheadedness upon standing)
- arrhythmias (interruption of impulse transmission from the atria to the ventricles—"heart AV nodal conduction block")

Therefore, they are contraindicated in people with preexisting bradycardia or arrhythmia. Other side effects are bronchoconstriction (restriction of airways) leading to asthma-like attacks, erectile dysfunction, chronic general fatigue (lethargy), and depression.

CONCLUSION

Beta blockers have been remarkably successful in the treatment of serious medical conditions such as coronary heart disease and hypertension. The prevalence of these conditions in contemporary society has resulted in the prescription of these drugs across a large segment of the population. The use of beta blockers as PEDs is limited to a few sports in which calmness and a steady hand are essential, but they are highly effective in these sports. WADA has banned them from archery and shooting in and out of competition, and from automobile sports, billiards, darts, golf, skiing, snowboarding, and underwater sports in competition.

Close-Up: Sometimes the Consequences of Doping Can Be Quite Severe

The Daily NK *is an online newspaper published in South Korea* (http://www.dailynk.com). *It focuses on issues relating to North Korea, its people, society, culture, and government. The following is an item it ran after a North Korean pistol shooter was disqualified for doping with beta blockers in the 2008 Beijing Summer Olympics.*[18]

North Korea is likely to severely punish one of its athletes after he tested positive for prohibited drugs at the Beijing Olympics and was stripped of the medals that he had earned.

The International Olympic Committee (IOC) announced publicly on the 15th (August) that North Korean shooter Kim Jong Su has been stripped of his silver and bronze medals, which he won in the 2008 Beijing Olympics men's 10-m pistol and 50-m air pistol, respectively, after he tested positive for the banned beta-blocker propranolol.

A Daily NK's reporter met with an affiliate from the North Korean economic department after the International Olympic committee's announcement in Shenyang, China. He emphasized that "Kim humiliated the country in the Olympics. It is a serious problem that we cannot overlook in the current atmosphere. Even the authorities are outraged by this sudden drug problem."

The reporter asked the question if Kim Jong Il (leader North Korea) has mentioned this issue, to which the affiliate replied "I am not sure about that, but North Korea is a target for a lot of criticism from the outside world related to counterfeiting and drugs. From this scandal, we have earned more negativity."

Another inside source from North Korea says "If you humiliate the country like Kim Jong Su did, you will be secretly dragged somewhere and suffer. Defectors in China who have heard the news say that the North Korean authorities will deal out a severe punishment."

When Kim Jong Su returns to North Korea, he will receive disciplinary punishment from the Korean Physical Culture and Sports Guidance Committee, the highest organization in the field of the sports. However, there is a high possibility of punishments such as imprisonment and labor training.

Defectors say that the North Korean soccer players who reached the World Cup quarterfinal game in 1966 were placed in political prison camps, for their questionable activities abroad, as soon as they returned to North Korea. Also, when the North Korean national soccer team was defeated in the 1994 World Cup, Kim Jong Il ordered "Do not let them go abroad for years."

Internationally, if an athlete is caught using prohibited drugs, his previous records will be thrown out and he will not be able to compete in international competition. Giving a judiciary punishment in a country for a case dealing with prohibited drug usage is rare.

Kim, who is thus the first North Korean Olympic medalist caught for using prohibited drugs in this Olympics, is from the Ministry of Physical Culture and Sports and is also known as the Bronze medalist in the male 50-m pistol shooting at the 2004 Athens Olympics.

Reproduced by permission of Daily NK

REFERENCES

1. Black J. Reflections on drug research. *Br J Pharmacol.* 2010;161:1204−1216.

2. Hackney AC. Stress and the neuroendocrine system: the role of exercise as a stressor and modifier of stress. *Expert Rev Endocrinol Metab.* 2006;1(6):783−792.

3. Frishman WH. Beta-adrenoceptor antagonists: new drugs and new indications. *N Engl J Med.* 1981;305:505−506.

4. Akbar S, Alorainy MS. The current status of beta blockers' use in the management of hypertension. *Saudi Med J.* 2014;35(11):1307−1317.

5. Ahlquist RP. A study of the adrenotropic receptors. *Am J Physiol.* 1948;153:586−600.

6. Frishman WH. Alpha and beta-adrenergic blocking drugs. In: Frishman WH, Sonnenblick EH, Sica DA, eds. *Cardiovascular Pharmacotherapeutics.* Second ed. New York: McGraw Hill; 2003:67−97.

7. Beta-adrenergic Blocking Agents. Drugs.com. ⟨https://www.drugs.com/drug-class/beta-adrenergic-blocking-agents.html⟩; 2017. Accessed 06.25.2017.

8. Frishman WH. Fifty years of beta-adrenergic blockade: a golden era in clinical medicine and molecular pharmacology (commentary). *Am J Med.* 2008;121:933−934.

9. The World Anti-Doping Code International Standard—Prohibited List 2017. World Anti-Doping Agency. ⟨https://www.wada-ama.org/sites/default/files/resources/files/2016-09-29_-_wada_prohibited_list_2017_eng_final.pdf⟩ 2017; 2016. Accessed 07.04.2017.

10. List of Doping Cases in Sport, Wikipedia; 2017. ⟨https://en.wikipedia.org/wiki/List_of_doping_cases_in_sport⟩ (accessed 07.10.2017).

11. Hackney AC. *Exercise, Sport, and Bioanalytical Chemistry: Principles and Practice.* New York, New York: Elsevier−RTI Press; 2016.

12. James IM, Griffith DN, Pearson RM, et al. Effect of oxprenolol on stage-fright in musicians. *Lancet.* 1977;2(8045):952−954.

13. Kruse P, Ladefoged J, Nielsen U, et al. Beta-blockade used in precision sports: effect on pistol shooting performance. *J Appl Physiol.* 1986;61(2):417−420.

14. Hopkins WG, Schabort EJ, Hawley JA. Reliability of power in physical performance tests. *Sports Med.* 2001;31(3):211−234.

15. Clarkson PM, Thompson HS. Drugs and sport: research findings and limitations. *Sports Med.* 1997;24(6):366−384.

16. Hawkins MN, Barnes Q, Purkayastha S, et al. The effects of aerobic fitness and beta1-adrenergic receptor blockade on cardiac work during dynamic exercise. *J Appl Physiol.* 2009;106(2):486−493.

17. Athan A N, Sampson UI. Coping with pre-competitive anxiety in sports competition. *Eur J Nat Appl Sci.* 2013;1(1):1−9.

18. North Korean Doping Medalist in Danger? DailyNK. August 8, 2008. ⟨https://www.dailynk.com/englis/read.php?cataId = nk00100&num = 3983/⟩. Accessed 06.26.2017.

CHAPTER *10*

Athlete Testing, Analytical Procedures, and Adverse Analytical Findings

To enforce the World Anti-Doping Code (see Chapter 1: Overview: Doping in Sport), the World Anti-Doping Agency (WADA) in cooperation with national antidoping agencies and sports-governing organizations established strict guidelines and procedures to ensure accurate and fair assessment of athletes' biological specimens. The collection, handling, and testing of samples are based upon scientific research, best clinical chemistry practices, and contemporary investigative standards. This chapter is an overview of the detailed process by which athletes are tested, how their biological specimens are dealt with, and what happens if specimens suggest an athlete is doping and using performance-enhancing drugs (PEDs).

WADA TESTING PROCEDURES

The instructions for testing procedures to detect substances on the WADA Prohibited List are highly organized, detailed, and somewhat daunting. The *Athlete Reference Guide to the 2015 World Anti-Doping Code*, which explains the athlete's responsibilities, runs to nearly 200 pages.[1] This level of detail and standardization is necessary to protect and ensure that the specimens gathered and tested are viable and attributable only to the athlete in question. Table 10.1, excerpted from the *Guide*, gives the athlete specific directions for sample collection and handling.

It is easy to see from Table 10.1 how an athlete or coach might miss a step inadvertently, but ignorance of the rules or accidental nonadherence is not an excuse acceptable to WADA. Any variance from procedure can lead to a sanction ranging from a reprimand to a lifetime ban. If the violation warrants a period of ineligibility from competition, the length of the period may vary depending upon the type of violation, the substance detected, and the circumstances of the individual case (e.g., whether the athlete is a repeat offender).[2]

Doping, Performance-Enhancing Drugs, and Hormones in Sport. DOI: https://doi.org/10.1016/B978-0-12-813442-9.00010-9

Table 10.1 Biological Sample Collection Procedure That an Athlete Must Follow[1]

The 11 steps of sample collection

Urine specimen

As an athlete;
1. You can be selected for doping control at any time and any place.
2. A Doping Control Officer (DOC) or chaperone will notify you that you have been selected for doping control. The DCO or chaperone will inform you of your rights and responsibilities. These include the right to have a representative present throughout the process. You will be asked to sign a form confirming that you have been selected for doping control.
3. You should report immediately to the doping control station. The DCO or chaperone may allow you to delay reporting to the station for an activity such as a news conference or to complete a training session. However, once you have been notified that you have been selected for doping control, the DCO or chaperone will accompany you until the completion of the sample collection process.
4. You will be given a choice of individually sealed collection vessels. You may select one. You should verify that the equipment is intact and has not been tampered with. You should, at all times, maintain control of the collection vessel.
5. During the sample provision, only you and the DCO or chaperone of the same gender are permitted in the washroom. You will be asked to wash your hands. You will then be asked to raise or lower your clothing so that the DCO or chaperone has an unobstructed view while you provide the sample.
6. The DCO shall ensure, in your full view, that you have provided the minimum required volume: 90 mL *(urine)*. If at first you are unable to provide 90 mL, you will be asked to provide more until that level is met.
7. You will be given a choice of individual sealed sample collection kits. Choose one. You should verify the equipment is intact and has not been tampered with. Open the kit. Confirm the sample code numbers on the bottles, the lids, and containers all match. Now, you are going to split the sample, pouring at least 30 mL into the B bottle and the remaining urine into the A bottle. You will be asked to leave a small amount in the collection vessel. The reason for this is so the DCO can measure its specific gravity. Pour the urine yourself unless you need help. In this instance, you will need to provide consent for your representative or the DCO to pour on your behalf.
8. Next, seal both the A and B bottles. You (or your representative) and the DCO should verify that the bottles are sealed properly.
9. The DCO is required to measure the sample's specific gravity.[a] If it does not meet certain requirements, you will be asked to provide another sample.
10. Completing the doping control form is next. On this form, you should provide information about any medication—prescription or nonprescription—or dietary supplements you have taken recently. This form is also the place to note any comments you may have regarding any part of the doping control process. You will be asked whether you consent to have your sample used anonymously for research once the analysis of doping control purposes is completed. You may say yes or no. Be absolutely certain everything is correct, including the sample code number. Make sure, too, that the laboratory copy of the form does not include any information that could identify you. You will be asked to sign the form. At the completion of collection, you will receive a copy of your doping control form.
11. The laboratory process is next. Your samples are packed for shipping by a secure process. Your samples will be sent to a WADA-accredited laboratory. When processing your samples, that lab will adhere to the International Standard for Laboratories, ensuring the chain of custody is maintained. Your A sample is analyzed. Your B sample is securely stored. It may be used to confirm an Adverse Analytical Finding from the A sample. The lab will report the results of your sample analysis to the responsible antidoping organization and to WADA.

Blood sample (This may be asked to be provided)

The same conditions that apply for urine sample collection also apply to the collection of blood samples with regard to notification, identification, escorting, and explanation of the procedure

[a]Specific gravity—the ratio of the density of a substance to the density of a standard substance, usually water.
Source: *From Athlete Reference Guide to the 2015 World Anti-Doping Code. World Anti-Doping Agency. <https://www.wada-ama.org/en/resources/education-and-prevention/athlete-reference-guide-to-2015-code-online-version>; 2015 (accessed 05.05.2017).*

The athlete has recourse, such as requesting analysis of the back-up sample (the B sample) collected at the same time, a hearing of inquiry, and an appeal of the sanction. The appeal process is typically done first with the specific national sporting organization but can ultimately be taken to the international court of the Tribunal Arbitral du Sport (TAS; Court of Arbitration for Sport) in Lausanne, Switzerland for final review and adjudication (see "What Happens When an Athlete Gets a Positive Test Result" in this chapter).

BIOSPECIMEN ANALYSIS TECHNIQUES

All substances on the WADA Prohibited List can be detected in a biological sample. The kind of assay for detection varies for different substances. Techniques vary in quality too. WADA mandates the use of the most precise, accurate, and reliable ones in its 34 accredited laboratories around the world. Certification and accreditation is a critical step to ensure that results are valid and reliable regardless of where specimens are collected.[3] WADA has no problem unaccrediting laboratories that are not in compliance with best practices, which happened to the Brazilian national doping laboratory a few weeks before the 2016 Rio Summer Olympics.

The principle bioanalytical techniques are described briefly below. Table 10.2 shows 10 of the most common categories of doping agents used by elite athletes, with examples of each in parentheses, sample type usually analyzed, and principle bioanalytical methods of detection.[4]

Liquid Chromatography With Tandem Mass Spectrometry

High-performance liquid chromatography (HPLC) with tandem mass spectrometry (MS) is a hyphenated technique that uses LC (liquid chromatography) for separation and MS for detection. MS identifies compounds by the atomic mass-to-charge ratio of the analyte molecule. A database (library) of the known mass spectra of several thousand chemical compounds is stored on a computer. MS is a definitive and accurate technique for separation of analytes.[5] Common variations use electrospray ionization (EI) and atmospheric pressure chemical ionization (APCI),[6-8] depending on the type of specimen and the chemical nature of the analyte.

Table 10.2 The Most Common Categories of Doping Agents Used by Elite Athletes, with Specific Detection Methods (potential PED analytes for detection are in parentheses)
Anabolic androgenic steroids (e.g., androstendiol, androstenedione, testosterone)   Urine: gas chromatography/mass spectrometry (GC/MS), liquid chromatography with tandem mass spectrometry (LC/MS/MS), gas chromatography/combustion/isotope ratio mass spectrometry (GC/C/IRMS)
Stimulants (e.g., cocaine, benzfetamine, methylphenidate)   Urine: LC/MS, LC/MS/MS, GC/MS
Diuretics and masking agents (e.g., desmopressin, acetazolamide, bumetanide)   Urine: GC/MS, LC/MS, LC/MS/MS
Glucocorticoids (e.g., cortisol)   Blood; LC/MS/MS
Peptide hormones and related agents (e.g., growth hormone, luteinizing hormone, erythropoietin)   Blood: enzyme-immunoassays, biomarkers (insulin-like growth factor 1, procollagen III peptide), LC/MS/MS, isoelectric focusing (IEF), sodium dodecyl sulfate/polyacrylamide gel electrophoresis (SDS/PAGE), hematocrit measurements (erythrocyte size and iron content)
Cannabinoids (e.g., THC, marijuana, hashish)   Urine or blood: GC/MS
Beta-2 agonists (e.g., salbutamol, formoterol, salmeterol)   Urine: GC/MS
Hormone/metabolic modulators (e.g., clomiphene, tamoxifen, testolactone)   Urine: GC/MS
Narcotics (e.g., morphine, oxycodone, methadone)   Urine: GC/MS
Beta blockers (e.g., acebutolol, labetalol, podolol)   Urine: GC/MS

Gas Chromatography/Mass Spectrometry

GC (Gas Chromatography)/MS uses a different separation technique. GC works on the principle that a mixture will separate into individual substances when heated into a gaseous state. The gases are carried through a column with an inert gas such as helium. As the separated substances emerge from the column opening, they flow into the mass spectrometer for detection as with LC/MS. GC/MS also is highly accurate.[7,8]

Gas Chromatography/Combustion/Isotope Ratio Mass Spectrometry

GC/C/IRMS ascertains the relative ratio of light stable isotopes of carbon ($^{13}C/^{12}C$), hydrogen ($^{2}H/^{1}H$), nitrogen ($^{15}N/^{14}N$), or oxygen ($^{18}O/^{16}O$) in individual chemical compounds separated from complex mixtures. The ratio of these isotopes in natural materials is different from that in synthetic materials; hence, GC/C/IRMS can determine whether an analyte is synthetic. The primary prerequisite is that the compounds of the sample mixture be amenable to GC, that is, suitably volatile and thermally stable. Some polar compounds require further chemical modification or transformation (derivatization); in such cases, the relative stable isotope ratio of the derivatization agent must also be determined.[9]

Enzyme-Linked Immunosorbent Assay

Enzyme-Linked Immunosorbent Assay (ELISA) is a microplate-based technique for detecting and quantifying substances such as peptides, proteins, antibodies, and hormones. The analyte is the antigen in this procedure, that is, the substance to be detected. The technique has other names, such as enzyme immunoassay (EIA). The antigen must be immobilized to a solid surface (the well of a plate) and then complexed with an antibody that is linked to an enzyme. Detection is accomplished by assessing the conjugated enzyme activity between the antigen and the antibody via incubation with a substrate to produce a measurable product. Because of the variety of chemical substrates available, detection can be quantified with chromogenic, chemifluorescent, and chemiluminescent imaging. The most crucial element is to have a highly specific antibody—antigen interaction.

ELISA can have several modifications. The key step, immobilization of the antigen of interest, can be accomplished by direct adsorption to the assay plate well or indirectly via a capture antibody that has been attached to the plate well. The antigen is detected either directly (primary antibody) or indirectly (secondary antibody). The sandwich ELISA format is considered the most sensitive and robust— "sandwich" because the analyte is bound between the capture antibody and the detection antibody.[10]

Gel Electrophoresis

Gel electrophoresis is a separation technique using electrical charge. It is used to separate mixtures of RNA, DNA, and protein structures according to molecular size and charge. The gel (e.g., agar or polyacrylamide) is the medium through which a charge is applied. The biospecimen is applied to the medium, and the electrical field causes molecules to move or separate into difference bands (i.e., the negatively charged end of the gel pushes the molecules, and the positively charged end pulls them). The larger molecules in the specimen move more slowly than the smaller ones, so that molecules of various molecular sizes separate into distinct bands in the gel medium. After the electrophoresis separation, the molecules in the gel are stained with chemical compounds to make them visible, the specific compound depending upon the nature of the analyte. Reference mixtures of all potential analyte molecules can be analyzed and compared to the specimen of interest. In the case of certain proteins, additional separation procedures may be used in concert with the electrophoresis separation, such as isoelectric focusing (which determines a molecule's magnitude of charge at a select pH) or SDS-PAGE (sodium dodecyl sulfate-polyacrylamide gel electrophoresis, which determines the analyte's molecular length and mass-to-charge ratio).[3,4,11]

STRENGTHS AND WEAKNESSES OF THE BIOANALYTICAL PROCEDURES

All bioanalytical chemistry procedures have the potential for error of a biological, technical, or human origin. Equipment can be out of calibration, chemical reagents can be expired, and individuals can do procedures incorrectly, leading to false positives and false negatives. The WADA accreditation process strives to ensure that errors are kept to an absolute minimum.[11] Nonetheless, they do occasionally occur, which is why a confirmation analysis is done.

An obvious requirement of the system is to have a sample of the target drug and an assay for it. A case in point is the Bay Area Laboratory Co-Operative (BALCO) scandal. The company, started by Victor Conte near San Francisco, supplied anabolic steroids to athletes. Tetrahydrogestrinone (THG, or "The Clear") was a designer steroid developed for Conte that affects the user like other steroids and testosterone derivatives (see Chapter 2: Anabolic Androgenic Steroids). It was undetectable because antidoping laboratories did not have it in their databases. Trevor Graham, a coach, anonymously mailed a sample of it to a representative of the US Anti-Doping Agency, who forwarded it to an analytical laboratory at University of California, Los Angeles (UCLA), which developed a detection procedure. Eventually, several well-known athletes tested positive for THG, and it tarnished their careers: track and field athletes Marion Jones, Tim Montgomery, and Dwain Chambers, and baseball star Barry Bonds, who had all been doping for years undetected.[12]

WAYS IN WHICH ATHLETES AVOID DETECTION

Athletes will go to extraordinary lengths to continue using PEDs and procedures while avoiding detection. Some of these means are ingenious, others are plain foolish.

Diuretics

Diuretics are widely prescribed ethical drugs that increase the rate of urine flow and sodium excretion to adjust the volume and composition of body fluids, notably to treat hypertension. There are several categories that vary in chemical structure and properties, effects on urinary composition and renal hemodynamics, and mechanism of action. Examples: acetazolamide, amiloride, bumetanide, canrenone, chlorthalidone, ethacrynic acid, furosemide, indapamide, metolazone, spironolactone, thiazides (bendroflumethiazide, chlorothiazide, hydrochlorothiazide), and triamterene. Diuretics are readily available over the counter, and drugs marketed for other purposes (e.g., caffeine to increase alertness) have diuretic effects. Doping athletes use them to increase urine excretion rate and thereby mask the use by increasing the elimination of a banned substances. Diuretics are routinely screened for by antidoping laboratories, and WADA bans them both in and outside of competition. Excessive diuretic use can result in compromised physical performance and health consequences (Fig. 10.1).[13]

Masking Agents

Some masking agents, though not all, can impair detection of banned substance in urine or blood, and for this reason, they too are banned by WADA. Plasma expanders work on the same principle as diuretics, dilution and elimination, but for blood instead of urine. By increasing the amount of plasma, the fluid component of blood, they make detection of the banned substance more difficult. Common plasma expanders are albumin, dextran, and hydroxyethyl starch. Some can be purchased over the counter at pharmacies. They can cause a dangerous rise in blood pressure.

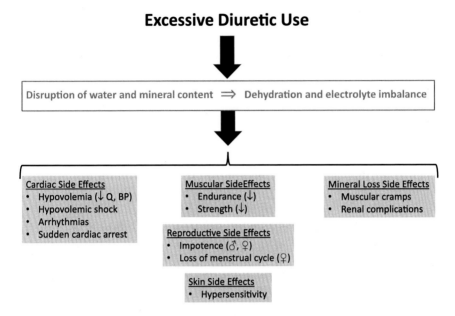

Figure 10.1 Major harmful side effects of excessive diuretic use. Q, cardiac output; BP, blood pressure.

Epitestosterone is used by athletes who take testosterone. It works by correcting the disturbed natural ratio between testosterone and epitestosterone (T/E ratio). The urine test for testosterone doping is based on this ratio. An athlete taking artificial testosterone will have an abnormally high ratio, which epitestosterone will return to normal, yielding a negative reading on the urine test.[4,13]

Other Methods

Athletes have tried less sophisticated tricks such as substituting a clean urine sample from another person. The sample collection and chain of custody procedures developed by WADA make success with substitution very unlikely. Also, analytical procedures are sophisticated enough to determine whether the sample came from the subject athlete or not. A well-known example of switching involved a fake penis device called the Whizzinator, which was found on Onterrio Smith, a former National Football League running back. The Whizzinator kit could be purchased complete with dried urine and syringe, heater packs (to keep the sample at body temperature), a false penis (available in several skin tones), and instruction manual. There was also a female version. The manufacture has since been brought to trial for attempting to defraud the government and conspiracy to sell drug paraphernalia.

Simple avoidance of testing is no longer an option either. Three missed tests in a period of 12 months, down from 18 months as of 2015, is counted the same as a failed test. To insure that athletes will be available for testing, they must provide to their national antidoping agencies a home address, training schedules and venues, competition schedule, regular personal activities schedule (e.g., school, work), and in most cases, a 60-minute window of time each day when they can be found.

WADA developed the online Anti-Doping Administrative Management System (ADAMS), which athletes use to indicate availability for out-of-competition testing during training periods. ADAMS has been implemented and is used by nearly 60 international sport federations, more than 40 national antidoping organizations, and all WADA-accredited laboratories. About 100,000 athlete profiles have been logged. Although ADAMS is not mandatory, WADA strongly recommends it. It allows antidoping agencies to coordinate activities among one another and to fulfill their responsibilities under the World Anti-Doping Code.[1,2]

WHAT HAPPENS WHEN AN ATHLETE GETS A POSITIVE TEST RESULT?

If an athlete's A sample is positive—an Adverse Analytical Finding—the athlete's national antidoping agency responsible for results management will conduct an initial review. In the United States, that would be the US Anti-Doping Agency. Initially, the review is focused on two questions:

- Did the athlete have a therapeutic use exemption (TUE) for the substance found in the sample?
- Was the sample collection and analysis done according to WADA procedures, or was there a departure from protocol?

The athlete is notified in writing of the positive result for the A sample and their rights regarding analysis of the B sample. If the athlete requests a B sample analysis, or if the antidoping organization does, the athlete may attend or send a representative. In the meantime, a provisional suspension is imposed.

The WADA Code gives the athlete the right to a hearing on the provisional suspension. If the B sample confirms the positive result of the A sample, the national antidoping agency will proceed with the results management process. If the B sample does not confirm the A sample, no further action is taken, and the provisional suspension is lifted. WADA documentation of the process, the *Results Management, Hearings and Decisions Guidelines*, is also an extensive document. WADA has worked diligently to standardize and document the procedures to protect the athlete's rights, and the course of action is well laid out. Fig. 10.2 is a flow diagram of the process.[1,2]

An antidoping rule violation normally results in disqualification, imposition of a period of ineligibility, mandatory publication of the violation, and financial sanctions to the athlete.

- Disqualification means that the athletic performance associated with the testing time is voided, any awarded medals are repealed, points awarded toward an overall seasonal tally are lost, and all prize money is forfeited. The athlete's results in other competitions in the same sporting event (e.g., the Olympic Games) may also be disqualified. Generally, disqualification is retroactive from the date of the rule violation (the date of collection of the positive sample) until the commencement of any provisional suspension or ineligibility period.
- Ineligibility means the athlete cannot take part in any competition or the activities of any international sporting organization, its national federations, or their sporting clubs. This is a very strict provision and includes training with the athlete's sports club or team or using their facilities. They cannot take part in any competition authorized or organized by signatories of the code (such as the International Olympic Committee, the International Paralympic Committee, the national-level Olympic committees), or any professional league, or any international or national-level event organization, or any elite or national-level sports activity funded by a governmental organization. The period of ineligibility is typically 4 years for intentional cheaters. It depends, however, upon the type of violation, the prohibited substance or method used, the nature of the athlete's conduct, and the athlete's degree of fault. Violations can also lead to no ineligibility if the athlete can establish no fault or negligence. Also, in some unique circumstances, a warning may be issued if the athlete's degree of fault is very low. This situation might apply where the substance was taken in a product, typically a dietary supplement, containing a prohibited substance that was not

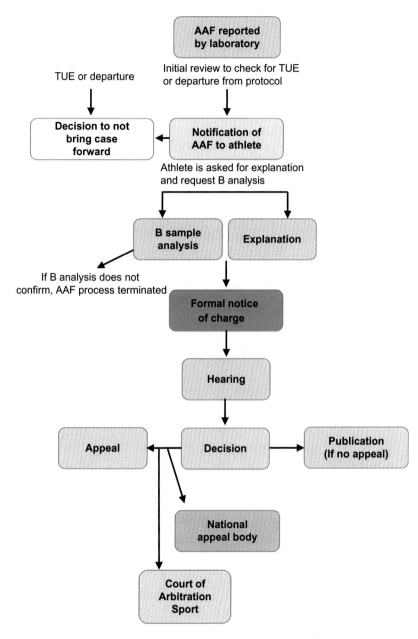

Figure 10.2 The steps following an Adverse Analytical Finding (AAF) in a drug test.[1]

disclosed on the product label or in an informational search. Allowances also are made when an athlete has demonstrated "collaboration and substantial assistance" to the antidoping agency. This is the cooperation of athletes and others who acknowledge their mistakes and are willing to step forward and bring antidoping violations to light. "Substantial assistance" means aiding an antidoping agency, the police or prosecuting authority, or a professional disciplinary body that results in the bringing of a new case against someone else, or discovering the possibility of doing so. This means there is full disclosure, in writing, of everything the athlete knows about doping by any person, including themselves, and full cooperation with the authorities. A period of ineligibility may be reduced by up to half if the athlete voluntarily admits doping before the agency files notice of a rules violation.

• Public disclosure of the violation is done as a future deterrent to athletes. All antidoping agencies must, except in the case of a minor offense, publish the name of the athlete, the nature of the violation, and the consequences, within 20 days after a final ruling. If the final decision is that no violation occurred, the decision may be disclosed only with the consent of the athlete.

There is a statute of limitations by which any proceeding must be brought within 10 years of the date of the alleged violation.

CONCLUSION

The collection and bioanalytical testing of athletes' biological specimens is a detailed and complex process. To standardize the practices and protect the athlete from being falsely accused of doping, WADA has established strict criteria for collection and analysis and the adjudication process if a specimen tests positive for a banned substance. These procedural safeguards are constructed to ensure that cheaters are caught and punished, and innocent athletes are not unfairly penalized. In an attempt to stay ahead of the dopers, WADA is constantly revising and fine tuning its procedures and striving to use state-of-the-art science in its methodology. Even though it can be daunting, it is the responsibility of the athlete and their coaching staff to know, understand, and stay current with the rules and policies as laid out annually by WADA.

Close-Up: Fox Guarding the Henhouse—The State-Sponsored Russian Doping Scandal

The monitoring of athletes for use of banned PEDs is a team effort. The number of athletes and sporting competitions worldwide is beyond the logistical scope of one organization. WADA relies on national sports organizations and antidoping agencies to help it fulfill its mission (see Chapter 1: Overview: Doping in Sport). Many of these organizations are responsible for some aspects of supporting and supervising the WADA-accredited antidoping laboratories that store and analyze the athletes' biological specimens.

In 2014, the Russian 800-m runner Yuliya Rusanova went on German television and accused the Russian sports system of large-scale systematic doping of their athletes. These claims were supported by her fiancé (now husband) Vitaly Stepanov, who worked for the Russian national antidoping agency. Russian sports organizations, politicians, and the government denied their claims, but later Dr. Grigory Rodchenkov, director of the Russian antidoping laboratory for the Sochi Olympics, corroborated them with his own story of illicit events.

To investigate the allegations, WADA retained Richard McLaren, a Canadian attorney. In 2016, McLaren released a preliminary and then a final report on state-sponsored doping in Russia. His investigative team found corroborating evidence of sponsored doping after conducting extensive witness interviews, reviewing thousands of documents, cyber analysis of hard drives, forensic analysis of urine sample collection bottles, and laboratory analysis of individual athletes' biological samples.[14] The final report concluded that the evidence shows "beyond a reasonable doubt" that Russia's Ministry of Sport, the Centre of Sports Preparation of the National Teams of Russia, the Federal Security Service (successor to the KGB), and the WADA-accredited laboratory in Moscow had "operated for the protection of doped Russian athletes" within a "state-directed failsafe system."[14]

There were substantial incriminating findings in the report. Some of the most damning were as follows: More than 1000 Russians from at least 30 sports benefited from the state-sponsored doping program between the years 2011 and 2015. Many of these athletes competed at the 2012 London Olympics, 2013 World Athletics Championship in Moscow, and the 2014 Winter Olympics in Sochi, and some were medalists. Emails were found asking for instructions from the Russian Ministry of Sport on what to do with a positive biosample—save or quarantine. Spreadsheets were found containing lists of athletes whose questionable samples had been saved rather than reported. A bank of clean urine samples were retained in Moscow for the use of athletes. Directions for formulating a cocktail of PEDs—known as the "Duchess"—with a very short detection window were developed.

McLaren's final report added considerable depth and supporting evidence to the preliminary findings that Russia operated a state-sponsored doping program. The reports were met with denials from Russia and calls for more proof from the International Olympic Committee and WADA. Based upon the findings, WADA recommended a blanket ban on all Russian athletes for the 2016 Rio Summer Olympics. The International Olympic Committee decided to allow Russian athletes to compete if they could convincingly show that they had not been doping prior to the games. This restriction resulted in over a third of the athletes being unable to complete.

This entire episode has called into question the wisdom of WADA for allowing national sports organizations and antidoping agencies to have such a large role in the monitoring of athletes. But the logistical requirements of monitoring a worldwide sports system are so enormous that WADA has been obliged to maintain the present operational model. Whether this remains the status quo in the future remained to be seen.

REFERENCES

1. Athlete Reference Guide to the 2015 World Anti-Doping Code. World Anti-Doping Agency. <https://www.wada-ama.org/en/resources/education-and-prevention/athlete-reference-guide-to-2015-code-online-version>; 2015 (accessed 05.05.2017).

2. At-a-Glance Series—World Anti-Doping Agency—Play True. <https://www.wada-ama.org/en/resources/education-and-prevention/at-a-glance-about-anti-doping>; 2016 (accessed 6.04.2017).

3. International Standards for Laboratories — World Anti-Doping Agency. <https://www.wada-ama.org/en/resources/laboratories/international-standard-for-laboratories-isl>; 2016 (accessed 04.06.2017).

4. Butch AW. Sport drug testing laboratories. *Clinical Laboratory News.* Jan. 2014 Available from: <https://www.aacc.org/publications/cln/articles/2014/january/sports-drug>; 2014 (accessed 04.14.2017).

5. Patel KN, Patel JK, Patel MP, et al. Introduction to hyphenated techniques and their applications in pharmacy. *Pharm Methods.* 2010;1(1):2−13.

6. De Hoffmann E, Charette J, Strooban V. *Mass Spectrometry. Principles and Applications.* New York: John Wiley and Sons; 1996:91−97.

7. Pitt JJ. Principles and applications of liquid chromatography−mass spectrometry in clinical biochemistry. *Clin Biochem Rev.* 2009;30(1):19−34.

8. Vogeser M, Parhofer KG. Liquid chromatography tandem−mass spectrometry (LC−MS/MS)—technique and applications in endocrinology. *Exp Clin Endocrinol Diabetes.* 2007;115(9):559−570.

9. Cawley AT, Flenker U. The application of carbon isotope ratio mass spectrometry to doping control. *J Mass Spectrom.* 2008;43(7):854−864.

10. Darwish IA. Immunoassay methods and their applications in pharmaceutical analysis: basic methodology and recent advances. *Int J Biomed Sci.* 2006;2(3):217−235.

11. Clinical and Laboratory Standards Institute. *Protocols for Determination of Limits of Detection and Limits of Quantitation, Approved Guideline.* Wayne, PA: CLSI; CLSI document EP17; 2004.

12. Fainaru-Wada M. *Game of Shadows: Barry Bonds, BALCO, and the Steroids Scandal that Rocked Professional Sports.* New York: Gotham Books; 2006.

13. Cadwallader AB, de la Torre X, Tieri A, et al. The abuse of diuretics as performance-enhancing drugs and masking agents in sport doping: pharmacology, toxicology and analysis. *Br J Pharmacol.* 2010;161(1):1−16.

14. BBC New Report (Online)—Russian Doping: McLaren Report Says more than 1,000 Athletes Implicated. December 2016. <http://www.bbc.com/sport/38261608>; 2016 (accessed 04.30.2017).

CHAPTER *11*

The Future of Performance Enhancement in Sport

For millennia, athletes have strived to improve their performance and best their competitors, primarily through physical training. In the modern era, it is not uncommon for some professionals to train 3 to 6 hours a day and competitive nonprofessionals 1 to 3 hours a day. The body adapts to this training stimulus by enhancing cellular, tissue, and organ systems to meet the requirements of improved performance.[1] These improvements in biological function have enabled continuing advances in performances, as reflected in new world records for sporting events. Fig. 11.1 shows an example, the large reduction in men's and women's 800-m swim time over the last 100 years or so (*N.B.*: The performances are assumed to be free of doping.).[2] Many other sporting events show similar dramatic improvement. The primary driver of this phenomenon is the increased level of training. Better health care, nutrition, and equipment are also contributing

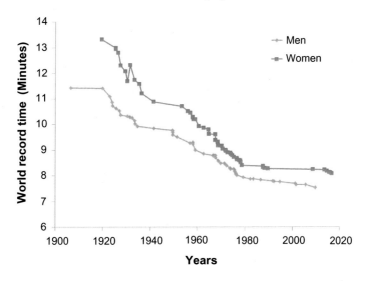

Figure 11.1 Changes in the world record for the 800-m freestyle swimming over the past 100 years.[2]

Doping, Performance-Enhancing Drugs, and Hormones in Sport. DOI: https://doi.org/10.1016/B978-0-12-813442-9.00011-0

factors, but the bottom line is that today's athletes are far more physically fit and trained and therefore have better developed biochemical and physiological systems for competition performance.

Financial compensation for top athletes is now phenomenal, and that tempts some to seek any mechanism to further improve performance. Doping while training is a quick way to improve performance faster and possibly to a greater extent. Better technology helps—shoes, for example—but not as much as better biology. Why then do athletes not just train more, without drugs, to improve their performance? Put simply, because doping may allow them to overcome a genetic limitation in their biology, or to train harder without becoming injured.[1]

CAN PREVENTION AND DETECTION DETER DOPING?

The World Anti-Doping Agency (WADA) and national antidoping agencies go to great lengths to inform and educate athletes and their coaches. These efforts focus on identifying banned substances clearly, setting forth the punishments for violations, and warning of the health consequences. The message is conveyed in several ways: print, online, on the airwaves, and direct messages delivered by representatives. Statistics on the number of testing procedures done and the number of positive outcomes for a given sport are publicized. But evidence for the degree of deterrence is less than clear, as is the information on those who doped but were never caught. It is widely assumed that the prevention and detection programs work, but nobody knows how well.[3-5]

EDUCATION OF ATHLETES: IS THAT ENOUGH?

To deter doping, WADA and national antidoping agencies strive to affect attitudes and actions of athletes by helping them move through the stages of behavioral change (Fig. 11.2).[6] Athletes are bombarded with educational materials and media stories about doping. Evidence indicates that they are generally knowledgeable about doping and its negative health consequences, if less familiar with procedural details of testing and which substances are on the WADA Prohibited List at a given time.[5] The question remains, however, to what degree is education changing athlete behavior? We simply do not know the answer to that question.

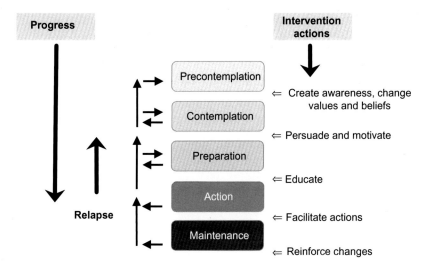

Figure 11.2 Stages of behavioral change in humans.[6]

When education fails to deter, the system is structured to punish chea-ters. How effective is that? The long-term outcome is likely to be best if punishment of the undesired behavior (doping) is accompanied by posi-tive reinforcement of the desired behavior.[3] But reinforcement of not dop-ing is logistically challenging for WADA and other agencies. Nonetheless, they need to give some creative thought to how it can be done.[4]

STAYING AHEAD OF THE DOPING ATHLETE: IS IT POSSIBLE?

Although doping is now a field of rigorous scientific study, that does not mean that dopers necessarily use unbiased methods and careful experimentation and come to factually based conclusions on which doping agent to use. They may consult research publications and online websites to choose a drug, but from that point on, there is no unbiased testing for safety and efficacy beyond the individual's own trial and error, which is compromised by the placebo effect. To sup-posedly aid the athlete, a large and highly developed black market and underground network exists for doping materials and information, much of it factually wrong, as unscrupulous vendors hype their wares.

Some highly trained scientists and physicians in the past as well as currently work with athletes in their doping practices. This aid may be well meant for safety reasons or cynically given for financial and other rewards. In any case, it is illegal, ethically ruinous, and potentially harmful to the athlete. Two examples:

- Dr. Patrick Arnold is an American organic chemist known for introducing androstenedione, 1-androstenediol, and methyl-hexanamine into the dietary supplement market, and for creating the designer steroid tetrahydrogestrinone (THG, "The Clear"). THG was at the heart of the Bay Area Laboratory Co-Operative (BALCO) professional sports doping scandal (see Chapter 10: Athlete Testing, Analytical Procedures, and Adverse Analytical Findings).
- About 10 years ago, Spanish police raids uncovered hundreds of bags of blood and plasma in the offices of former cycling sports medicine doctor Eufemiano Fuentes ("Dr. Blood," as Tyler Hamilton, a professional cyclist and one of his former clients, called him). Several years later, Fuentes was implicated again as part of *Operación Galgo* (Operation Greyhound). This was a series of simultaneous raids across Spain where police seized a large quantity of anabolic steroids, hormones, erythropoietin, and laboratory equipment for blood transfusions.[7]

Both of these individuals were arrested, convicted, and imprisoned, but no one knows how prevalent their practices are and how many other scientists and physicians might be helping athletes stay one step ahead of the antidoping regulators.

GENETIC DOPING: IS IT THE FUTURE?

Gene doping is an outgrowth of the medical field of gene therapy. But instead of introducing DNA into the body to restore a physiological function due to a damaged or missing gene, gene doping does it to gain an athletic performance edge.[8] WADA defines gene doping as the "nontherapeutic use of cells, genes, genetic elements, or modulation of gene expression, having the capacity to enhance performance."[9]

Gene doping clearly has performance-enhancing drug potential, and athletes are aware of it.[8-10] Dr. H. Lee Sweeney, professor of physiology at the University of Pennsylvania School of Medicine and a leading researcher in gene therapy, has been inundated with gene doping requests.[11] He is famous for his discovery in animals of a way to reverse muscle degeneration caused by diseases such as Duchenne muscular dystrophy, a sex-linked genetic disorder. In this disease, a critical muscle protein called dystrophin gradually becomes dysfunctional, leading to a loss of muscle fiber, an increase in fibrosis, and eventually complete loss of muscle function. Sweeney and his colleagues observed in normal mice that when the hormone insulin-like growth factor 1 (IGF-1) interacted with muscle fibers, it caused the cells to grow and become stronger. Inserting the gene that encodes IGF-1 into Duchenne mouse muscle cells produced the same adaptive effects.[12] Introduction of the IGF-1 gene produced what became known as "Schwarzenegger mice" because of their enhanced muscle development and functionality, obviously a potential benefit for sporting activities.

Dr. Ronald Evans of the Salk Institute in La Jolla, California, demonstrated that injecting mice with the gene that encodes the fat metabolism signaling protein mechanism (PPAR-δ; peroxisome proliferator-activated receptors delta) enabled them to run twice the distance of their nontreated littermates and increased the number of Type I muscle fibers (slow twitch endurance).[13] These genetically engineering animals were dubbed "marathon mice"—again, a potential benefit to some athletes.

Gene therapy, and by extension gene doping, are developing rapidly. Limited human trials on medical conditions have been highly promising.[14,15] Potentially dangerous side effects are not yet known and need to be ascertained by long-term clinical trials. Nonetheless, history suggests that gene doping will be the performance-enhancing procedure (i.e., drug) of the future for athletes inclined to cheat (if it has not already been used!).[8–10,15]

SHOULD DOPING BE LEGALIZED AND REGULATED?

There are those who have suggested that doping be legalized, controlled, and medically supervised. They include athletes, coaches, sports writers, scientists, physicians, and even some ethicists. While the prevalence of this opinion is difficult to quantify, it appears to be a minority one, especially in sports-governing organizations. Nonetheless, these proponents have been vocal and persistent over the decades since organized antidoping prevention and detection programs began. One prominent advocate is Dr. Julian Savulescu of the

University of Oxford in the United Kingdom. Writing in the *British Journal of Sports Medicine*, he argued that doping should be allowed and that the risk to an athlete's welfare is the only legitimate grounds for restriction.[16] Some points he and other advocates have made:[4,16]

- The current testing system has largely failed to deter and to catch cheaters. The probability of an athlete being detected for doping may be as low as 2.9%.
- Athletes may be approaching the limits of human performance, and the urge to excel makes doping inevitable.
- Legalization would allow structured regulation and lessen the health risk.
- The concept of a fair and level playing field is a myth, and the idea of an unfair advantage by doping ignores the realities of modern sport. Legal practices and devices (altitude chamber, nutritional practices, antiinflammatory drugs, etc.) improve performance and are not only allowed but encouraged. That is, many competitions are decided by who has the most money and resources, so why not allow the best medical doping ingenuity?

Ultimately, regardless of where they stand, most sports experts agree that the current testing system is too underfunded and intermittent in assessment to meet its goal of promoting drug-free sport.[4,16] Legalization is likely to remain an issue of heated discussion for the foreseeable future.

BINARY GENDER CONSTRUCTS AND THE FUTURE

Sport competition at the local, national, and international level is and nearly always has been segregated by biological sex: male and female. Biological sex is assigned at birth based upon reproductive organs. Whether it be little league baseball, youth football, National Collegiate Athletic Association (NCAA) sports, or the Olympics, there are few exceptions to the rule that boys compete against boys, girl versus girls, men versus men, and women versus women. This construct has existed for centuries and was established out of social and culture norms centered on the concepts of fair play and a level playing field. There is certainly a degree of sexism to that structure, as most of the leaders of organized sport during its development were men who were certain that "the fairer sex" could not physically compete with males.[17] We now know that this perception is not true in all cases. For example, at the 2016 Summer Olympics in Rio de Janeiro, Shelly-Ann Fraser-Pryce of Jamaica won the women's 100-m sprint in 10.71 seconds, faster than 25 ($\sim$20%) of the men in the 100-m competition at the same Olympics.[18] Also, importantly, Fraser-Pryce passed her International Olympic Committee−WADA doping tests at that competition.

New paradigms are emerging as science, culture, and societies are beginning to recognize and accept that the binary approach to sex and gender may not be the most correct or only acceptable model. Fig. 11.3 summarizes some aspects of the new perception, which raises challenging situations for sports-governing organizations and, in turn, antidoping agencies. Is the old paradigm of men's and women's categories for competition and record keeping appropriate? What about intersex individuals (see the close-up at the end of this chapter)? What about lesbian, gay, bisexual, transgender, and queer individuals? How are they to be treated in sporting competition? If fair play is something that we hold dear and fundamental, should the field be open to all? Respect and acknowledgment of human worth is as vital in the sporting context as in any social setting, and is taken to be a fundamental human right.[19,20]

To that end, the International Olympic Committee Medical and Scientific Commission convened a meeting in Lausanne, Switzerland in late 2015 to discuss issues of transgender athletes and updating the 2004 Stockholm Consensus policy recommendations on how and when transgender athletes could compete (Table 11.1). The 2015 meeting updated the guidelines and made them less onerous on transathletes but left them at the discretion of the various international sports-governing organizations for the 2016 Rio Olympics. Still, no transgender athletes were known to have competed at the 2016 Rio games.[20]

WHAT'S NEXT?

The future of sports doping is unclear. Antidoping agencies recognize a need to recalibrate the process, but exactly what that recalibration would involve is in question. Societies and nations must decide whether they want drug-free sport or legalized doping. These are difficult issues to struggle with as individuals and as a people. Sport is an integral part of the fabric of so many countries and indeed civilizations that the problem of doping, as it stands now, cannot and should not be

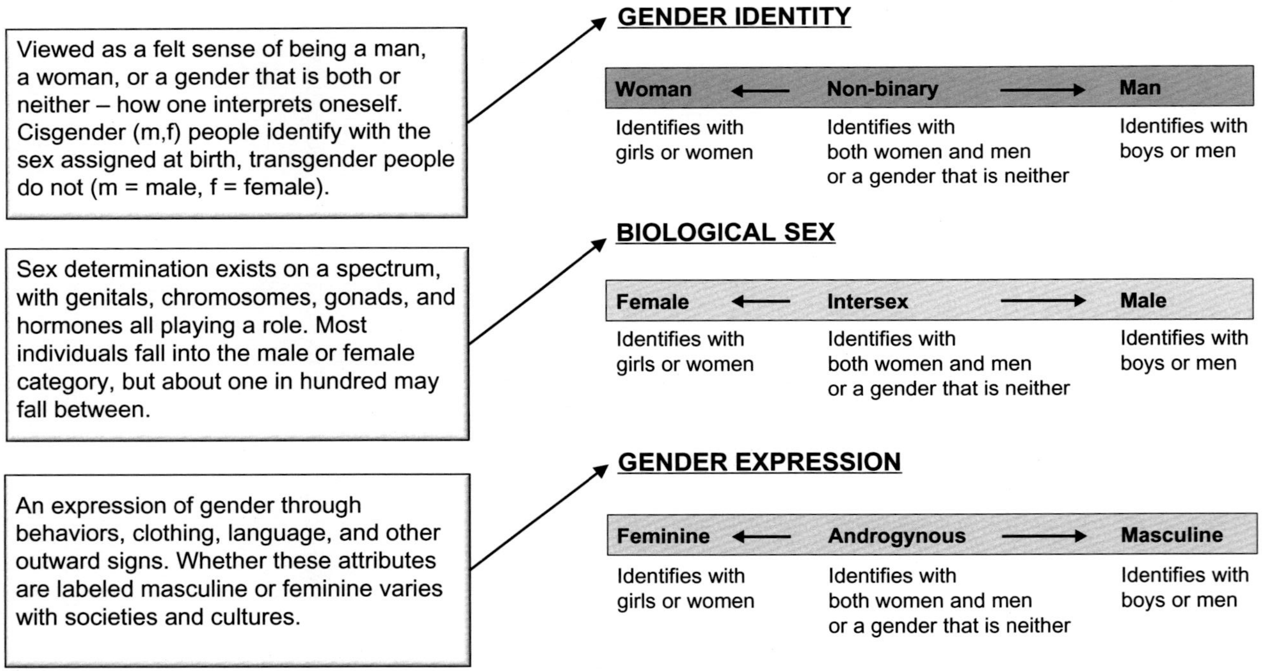

GENDER IDENTITY

Viewed as a felt sense of being a man, a woman, or a gender that is both or neither – how one interprets oneself. Cisgender (m,f) people identify with the sex assigned at birth, transgender people do not (m = male, f = female).

Woman	Non-binary	Man
Identifies with girls or women	Identifies with both women and men or a gender that is neither	Identifies with boys or men

BIOLOGICAL SEX

Sex determination exists on a spectrum, with genitals, chromosomes, gonads, and hormones all playing a role. Most individuals fall into the male or female category, but about one in hundred may fall between.

Female	Intersex	Male
Identifies with girls or women	Identifies with both women and men or a gender that is neither	Identifies with boys or men

GENDER EXPRESSION

An expression of gender through behaviors, clothing, language, and other outward signs. Whether these attributes are labeled masculine or feminine varies with societies and cultures.

Feminine	Androgynous	Masculine
Identifies with girls or women	Identifies with both women and men or a gender that is neither	Identifies with boys or men

Figure 11.3 Concepts of gender identity, biological sex, and gender expression.[20–22]

Table 11.1 The 2004 Recommendations Adopted by the International Olympic Committee Executive Board on How to Integrate Transgender Individuals into Competitive Sport[20]

The "Stockholm Consensus" recommendations called for inclusion of male–female and female–male transgender athletes so long as they met the following criteria:
- Gonadectomy
- Completion of anatomic changes consistent with their professed gender
- 2-year period of hormonal therapy "appropriate for assigned sex" and "in a verifiable manner"
- Legal recognition of their reassigned gender/sex

ignored, as that would have severe consequences for all who are interested, involved, and dedicated to sporting activities.

CONCLUSION

Doping in sport is a complex and emotional issue. This is true for those who abhor the practice as well as those who embrace it as a natural extension of our right to evolve and better ourselves. Regardless of where one lies on the continuum of belief, doping is now socially unacceptable, banned, and in some cases illegal. Whether this attitude will persistent is unclear, but it is the status quo. Those who choose to practice doping are playing with fire. They run the risk of not only getting metaphorically burned but of severely compromising the quality of their life and perhaps losing it.

Close-Up: Women Athletes—Taller, Faster, Stronger, and Hyperandrogenism

The issues of sex and gender have been paramount in the case of the South African track and field athlete Caster Semenya (Fig. 11.4) over the last few years. Semenya is biologically a woman but has masculine traits and meets the definition of being intersex (Table 11.2). The International Association of Athletics Federations (IAAF), the governing body for track and field, had a policy to exclude women athletes from competing as women if they were hyperandrogenistic or had related hormonal conditions—that is, had an excessive level of naturally produced testosterone—on the grounds that the condition confers an unfair advantage. Hormonal level is one of the factors by which an individual can be classified as intersex.

The IAAF had stated that testosterone is linked to lean body mass, so it influences strength, speed, and power, and hence gives hyperandrogenic women an advantage. The IAAF and the International Olympic

Figure 11.4 Caster Semenya at the 2011 Bislett Games.[23] Photo: Chell Hill.

Table 11.2 Definition of Intersex[19]

The United Nations Commission on Human Rights defines intersex as follows:
"Intersex people are born with any of several variations in sex characteristics including chromosomes, gonads, sex hormones, or genitals that do not fit the typical definitions for male or female bodies. Such variations may involve genital ambiguity, and combinations of chromosomal genotype and sexual phenotype other than XY-male and XX-female."[19]

Intersex traits are typically described medically as "disorders of sex development." This terminology, however, has been controversial since its introduction as people have experienced a negative impact of the use of the label.[22]

Committee set a permissible upper limit of 10 nmol/L testosterone for what constituted a woman, based on a research study of women in the World Championship competitions between 2011 and 2013. Semenya was forced to undergo sex testing to prove she was a woman by the biological definition. However, a 2014 study of hormonal profiles in elite athletes found that ~14% of women had high levels of testosterone and ~17% of men had low levels in comparison to the 10 nmol/L cut-point.

The researchers in this study concluded that there is "complete overlap between the sexes" in testosterone, and that the IAAF definition of a woman based on testosterone level was untenable.[24]

In 2014, Dutee Chand, a hyperandrogenic woman sprinter from India, was barred by the IAAF from competing against other female runners. She appealed the ruling and asked for reinstatement based upon the recent testosterone research. In 2015, the Court of Arbitration for Sport (see Chapter 10: Athlete Testing, Analytical Procedures, and Adverse Analytical Findings) suspended the IAAF ban, thus reinstating Chand's right to compete. This action resulted in the suspension of the IAAF criteria of hyperandrogenism and was one of the steps that ultimately allowed Caster Semenya to be eligible to compete in the 2016 Rio Olympics, where she won the gold medal in the women's 800-m. Not all Olympic women competitors were happy about this situation, and controversy persists. Conversely, some bioethicists and gender equality advocates argue that preventing women with high levels of testosterone or other intersex traits from participating is a form of discrimination, penalizing the athlete for a natural trait of her body. This issue will be one of continued discussion as part of the evolving social dialog on sex, gender, and sport.

REFERENCES

1. Hackney AC. *Exercise, Sport, and Bioanalytical Chemistry: Principles and Practice.* New York, New York: Elsevier—RTI Press; 2016.

2. World Record Progression 800 metres Freestyle. <https://en.wikipedia.org/wiki/World_record_progression_800_metres_freestyle>; 2017. Accessed 07.07.17.

3. Bower LD, Paternoster R. Inhibiting doping in sports: deterrence is necessary, but not sufficient. *Sport, Ethics Philos.* 2017;11(1):132—151.

4. Devine JW. Doping is a threat to sporting excellence. *Br J Sports Med.* 2011;45:637—639.

5. Alaranta A, Alaranta H, Holmila J, et al. Self-reported attitudes of elite athletes towards doping: differences between type of sport. *Int J Sports Med.* 2006;27:842—846.

6. Prochaska JO, DiClemente CC. Stages and processes of self-change of smoking: toward an integrative model of change. *J Consult Clin Psychol.* 1983;51(3):390—395.

7. Magnay J. Spanish doctor Eufemiano Fuentes told he does not have reveal athletes he treated in doping inquiry. *The Telegraph.* January 2013. <http://www.telegraph.co.uk/sport/othersports/cycling/9836906/Spanish-doctor-Eufemiano-Fuentes-told-he-does-not-have-reveal-athletes-he-treated-in-doping-inquiry.html>; 2013 Accessed 10.07.17.

8. Azzazy HM. Gene doping. *Handbook Exp Pharmacol.* 2010;195:485—512.

9. World Anti-Doping Agency. *Gene Doping. Play True.* 2005;1:1—13.

10. Murray T. An Olympic tail? *Nat Rev Genet.* 2003;4:494. Available from: https://doi.org/10.1038/nrg1135.

11. Brownlee C. Gene doping: will athletes go for the ultimate high? *Sci News.* October 26, 2004. https://www.sciencenews.org/article/gene-doping; 2004. Accessed 06.06.2017.

12. Barton ER, Morris L, Musaro A, et al. Muscle-specific expression of insulin-like growth factor I counters muscle decline in mdx mice. *J Cell Biol.* 2002;157(1):137−148.

13. Wang YX, Zhang CL, Yu RT, et al. Regulation of muscle fiber type and running endurance by PPAR-delta. *PLoS Biol.* 2004;2(10):e294.

14. Learn   Genetics.   <http://learn.genetics.utah.edu/content/genetherapy/success/>;   2017. Accessed 07.01.2017.

15. Skipper M. Gene doping: a new threat for the Olympics? *Nat Rev Genet.* 2004;5:720. Available from: https://doi.org/10.1038/nrg1461.

16. Savulescu J, Foddy B, Clayton M. Why we should allow performance enhancing drugs in sport. *Br J Sports Med.* 2004;38(6):666−670.

17. Messner MA. Sports and male domination: the female athlete as contested ideological terrain. *Sociol Sport J.* 1988;5:197−211.

18. Results Book: Rio 2016. Organising Committee for the Olympic and Paralympic Games in Rio 2016. Jeux olympiques d'été. Comité d'organisation. 31, 2016, Rio de Janeiro, Rio; 2016.

19. United Nations Office of the High Commissioner for Human Rights "Free & Equal Campaign Fact Sheet: Intersex" (PDF). Available from: <https://unfe.org/system/unfe-65-Intersex_Factsheet_ENGLISH.pdf>; 2015. Accessed 07.09.2017.

20. Genel M. Transgender athletes: how can they be accommodated? *Curr Sports Med Rep.* 2017;16(1):12−13.

21. Henig RM. Rethinking gender. *Natl Geogr Mag.* 2017;231(1):48−73.

22. Houk CP, Hughes IA, Ahmed SF, et al. Writing committee for the International Intersex Consensus Conference participants—summary of consensus statement on intersex disorders and their management. *Pediatrics.* 2006;118(2):753−757.

23. Wikimedia Commons. Caster Semenya at the 2011 Bislett Games. <https://commons.wikimedia.org/wiki/File:2011-06-09_Semenya.jpg>; 2011. Accessed 08.29.17.

24. Bermon S, Garnier PY, Hirschberg AL, et al. Serum androgen levels in elite female athletes. *J Clin Endocrinol Metab.* 2014;99(11):4328−4335.

Note: Page numbers followed by "*f*," "*t*," and "*b*" refer to, figures, tables, and boxes, respectively.

A

Acromegaly, 59
Action potential, 94–95
Addison's disease, 37
Adenosine triphosphate (ATP), 45, 52–53, 67, 106
Adenylyl cyclase, 106
Adrenal glands, 37
Adrenergic receptors, 66, 68
  subtypes, 69*f*
Adrenocorticotropic hormone (ACTH), 37–40
Adverse Analytical Finding (AAF), 121, 123*f*
Age-related sarcopenia, 83
Agovirin, 15*t*
Agrovirin, 15*t*
Airway narrowing, 73
Albuterol, 66, 66*t*
Alpha-1 receptors, 69*f*, 106–107
Alpha-2 receptors, 69*f*, 106–107
American Civil War, 96
Amphetamine, 25–28, 30–31
  neurochemistry, 28
  sympathomimetic action, 32–34
Anabolic androgenic steroids. *See* Anabolic steroids
Anabolic steroids, 2–3, 5–6, 13, 21, 78–79, 116*t*
  early paradoxes and later confirmations in research findings, 18–19
  effects of abuse, reversible, and nonreversible, 19–20, 19*t*
  gender-specific effects, 20
  intracellular biochemical and nuclear actions, 16–18
    anti-catabolic actions, 17–18
    psychological effects, 18
    steroid hormone receptors, 16–17
  muscle mass, increasing, 14–16
    banned substances, 16
    prevalence of usage, 15–16
  popular agents, 15*t*
  treatment of endocrine disruptions, 13–14
Anadrol, 15*t*
Analgesia, 91

Anapolon, 15*t*
Anastrozole, 84
Anavar, 15*t*
Andarine (S4), 78*t*
Andro LA, 15*t*
Androgen receptors (ARs), 16–17, 80
Androgen therapy, 13
Androgenic hormones, 16–17
Androgenic–anabolic drugs, 78–79
Android, 15*t*
Andronate, 15*t*
Andropository, 15*t*
Androstenedione, 21
Andrusol-P, 15*t*
Anti-cholinergics, 66–67
Anti-Doping Administrative Management System (ADAMS), 121
Anti-doping agencies, 16, 121, 124
Anti-doping rule violation, 113, 122–124
Anti-Heroin Act of 1924, 96
Arformoterol, 66*t*
Aromatase, 80
Aromatase inhibitors, 84
Arrhythmia, 108–109
Asoprisnil (J867), 78*t*
Asthma drugs, 66–67
Asthma prevention and treatment, 65–67
Asthmatic athletes, use of inhalers by, 73*b*
Asthmatics, 71
Athlete Biological Passport (ABP), 58
Averbol, 15*t*
Avoidance of testing, 119–121

B

Banned substances, 16, 119
Bay Area Laboratory Co-Operative (BALCO), 53, 118–119
Behavior modification, 130, 131*f*
Beta blockers, 103, 116*t*
  blocking the actions of neurotransmitters and hormones, 106–107, 107*t*
  cardiovascular modulation and health, 103–104
  from lethargy to impotence, 108–109

Beta blockers (*Continued*)
  staying calm, cool, and collected on
    battlefield of sport, 105
  strong evidence of performance
    enhancement, 107–108
Beta-1 receptors, 69*f*, 104, 106–107
Beta-2 adrenergic receptor, 66
Beta-2 agonists, 65, 116*t*
  adrenergic and anabolic actions and
    reactions, 68–71
  for asthma prevention and treatment, 65–67
  for asthmatics, 71
  for breathing more and building muscle,
    67–68
  tachycardia, arrhythmias, syncope, 72
Beta-2 receptors, 68–71, 69*f*, 106–107
Beta-adrenergic receptor blockers, 103
Binary gender constructs, 134–135
Biological sex, concept of, 136*f*
Biospecimen analysis techniques, 115–118
  enzyme-immunoassay (EIA), 117
  enzyme-linked immunosorbent assay
    (ELISA), 117
  gas chromatography/combustion/isotope
    ratio mass spectrometry, 116–117
  gas chromatography/mass spectrometry, 116
  gel electrophoresis, 117–118
  liquid chromatography with tandem mass
    spectrometry, 115
  strengths and weaknesses of, 118–119
Blocadren (timolol), 105*t*
Blood doping, 53
BMS-564,929, 78*t*
Boldenone, 15*t*
Bradycardia, 109
Brevibloc (esmolol), 105*t*
Bronchoconstriction, 109
Bronchodilation (BD), 65, 67, 73–74
Bronchospasm, 72–75
Bystolic (nebivolol), 104, 105*t*

C
Cachexia, 14
Cadaver-derived growth hormone, 49–51
Caffeine, 26–32, 29*f*
cAMP-dependent protein kinase (PKA), 106
Cannabinoids, 116*t*
Cardioprotection, 103
Catecholamines, 30, 56, 69*f*, 91, 103
Cellular receptors
  actions of agonist and antagonist drugs on,
    77*f*
Chaperones, 40

Cheating, 7, 11
Chemical-based communication signaling, 28
Chronic kidney disease, 52
Chronic obstructive pulmonary disease
    (COPD), 65
Clenbuterol, 71–72
Clomiphene (Clomid), 78*t*, 79–80, 84
Cocaine, 31, 91
Controlled Substances Act of 1970, 91–92
Coreg (carvedilol), 104, 105*t*
Corgard (nadolol), 105*t*
Corticotropin-releasing hormone, 39–40
Cortisol, 17–18, 37–40, 82*t*
Cortisone, 39
Creutzfeldt–Jacob disease, 49–51
Cushing's syndrome, 37–38, 41–42
Cyclic adenosine monophosphate (cAMP),
    106–107
Cytokines, 41–43

D
Danabol, 15*t*
Danatrol, 15*t*
Danazol, 15*t*
Danoval, 15*t*
Dehydroepiandrosterone (DHEA), 21
Delatestryl, 15*t*
Delta (δ) opioid receptor, 95*t*
Depotestosterone, 15*t*
Desoxymethyltestosterone, 15*t*
Dexamethasone, 38*t*
DHEA-sulfate (DHEA-S), 21
Dianabol, 15*t*
Dichloroisoproterenol, 103–104
Disqualification, 122
Diuresis, 31–32
Diuretics, 116*t*, 119, 120*f*
Dopamine, 28, 95
Doping agents, 116*t*
Doping cases in the Summer and Winter
    Olympic, 4*t*
Doping control, 11
Doping in sport, 1
  consequences of, 109*b*
  deterrence, 130
  ethical issues of usage, sportsmanship,
    character, 6–8
    means-to-an-end philosophy, 7
    rewards, 7
  history and current state of the problem,
    1–2
  legalization and regulation of,
    133–134

performance-enhancing drugs (PEDs), usage of, 8–9
prevalence by sport, country, level of competition, 2–3
world anti-doping agency and legal problems, 3–6
Doping while training, 130
Drug dependence, 97
Duchenne muscular dystrophy, 83
Durathate, 15t
Dystrophin, 132–133

E
Education of athletes, 130–131
Electrical polarity, 94–95
Emotional/physical stress, 68
Endocrine disruptions, treatment of, 13–14
Endocrine function, 81
Endorphins, 95–96
Endurance athletes, 18–19
Enobosarm (Ostarine), 78t, 80–81
Enzyme-immunoassay (EIA), 117
Enzyme-linked immunosorbent assay (ELISA), 117
Ephedrine, 25–28
Epinephrine, 65, 68, 91, 103, 106
Epitestosterone, 120
Equipoise (veterinary), 15t
Ergogenic PEDs, 39
Erythropoiesis, 51
promoting, 18–19
Erythropoiesis-stimulating agents (ESAs), 52–53, 61
Erythropoietin (EPO), 51–56, 59
actions of, 57–58
increasing hemoglobin and erythrocyte levels, 55f
Essential fat, 33
Estradiol-beta-17, 81
Estrogen, 82t, 84–85
Estrogen receptor (ER), 77, 79, 86
Estrogen replacement therapy, 14
Euphoria, 95
Everone, 15t
Exemestane, 84
Exercise as treatment for breast cancer survivors, 86b
Exercise training, 54, 92, 98–100

F
Fat-free mass (FFM), 33, 34f, 52
Federal Food, Drug, and Cosmetic Act of 1938, 91–92

Fenoterol, 66
Fentanyl, 93
Financial rewards, 7
Finaplix, 15t
First World Conference on Doping in Sport, in Lausanne, Switzerland, 2
Fluoxymesterone, 15t
Food and Drug Administration, 26
Formoterol, 66t
FOXO catabolic pathway inhibition, 68–71
Free fatty acid (FFA) mobilization, 43–44

G
γ-aminobutyric acid (GABA), 95
Gas chromatography/combustion/isotope ratio mass spectrometry, 116–117
Gas chromatography/mass spectrometry (GC/MS), 116
Gel electrophoresis, 117–118
Genabol, 15t
Gender expression, concept of, 136f
Gender identity, concept of, 136f
Gene doping, 132–133
Gene therapy, 132–133
Glucagon, 56
Glucocorticoid receptors (GRs), 17–18, 40–43
Glucocorticoids, 37, 116t
anti-inflammatory and pulmonary actions, 37–39
endocrine and immune actions, 39–43
and ethical dilemma in sports medicine, 46b
functions of, 43–44
helping athletes recover faster, breathe better, and burn more fat, 39
inflammatory cascade involving, 42f
pseudo-Cushing's syndrome and mineralocorticoid actions, 44–45
Glucose, excessive, 72
G-protein receptors, 21
Greek Olympic athletes, 10
Growth hormone, 49–52, 54
acromegaly, 59
actions of, 56–57
regulation of, 54f

H
Hallucinogens, 91
Halotestin, 15t
Harrison Narcotics Tax Act, 96
Heart muscle contraction, 103
Hematopoiesis, 51
Hemoglobin (Hb), 52–53, 58
Heroin, 91–92, 96

High-density lipoprotein cholesterol (HDL), 20
High-performance liquid chromatography with
    tandem mass spectrometry, 115
Histamine, 94–95
Hormone and metabolic modulators, 77
    androgen and estrogen modulation, in
        health and disease, 77–79
    contradictions and ambiguous evidence,
        83–85
        SARMs, 83–84
        SERMs, 84–85
    gender-based hormones, balance between,
        79–81
    selective receptor modulation drugs, 81–82
    thrombosis, embolisms, hot flashes, and
        hyperandrogenism, 85
        SARMs, 85
        SERMs, 85
Hormone replacement therapy, 14
Hormone/metabolic modulators, 116t
Hydrocodone, 91–92
Hydrocortisone, 38, 38t
11β-Hydroxysteroid dehydrogenase, 40
Hyperandrogenism, 137b
Hyperglycemia, 72
Hypothalamic–pituitary–adrenal (HPA) axis,
    39–40
Hypothalamic–pituitary–gonadal axis, 80

I

Indacaterol, 66t
Inderal and Inderal LA (propranolol),
    103–104, 105t
Ineligibility, 122–124
Inflammatory bowel disorders, 52
Inhaled corticosteroids, 66–67
Inhaled steroid drugs, 74–75
Inhaled β2-agonists, 73
Inhalers, use of, 73b
Insulin-like growth factor 1 (IGF-1), 16–17,
    54, 57, 132–133
    regulation of, 54f
International Association of Athletics
    Federation (IAAF), 5–6, 137–139
International Olympic Committee (IOC), 1–2,
    5–6
    ban on steroids, 16
Intersex, definition of, 138t
Isoelectric focusing, 117–118
Isoproterenol, 65–66

J

Janus kinase 2 signaling pathway, 54–56

K

Kappa (κ) opioid receptor, 95t
Kerlone (betaxolol), 105t

L

Lasofoxifene, 78t
Laudanum, 91–92
Lean body mass (LBM), 33, 34f, 52
Legalization, 134
Letrozole, 80
Levalbuterol, 66t
Levatol (penbutolol), 105t
LGD-4033 (Ligandrol), 78t
Lipolysis, 28–30
Lipolytic hormones, 56
Lipoprotein lipase (LPL), 20
Liquid chromatography (LC) with tandem
    mass spectrometry, 115
Lonavar, 15t
Long-acting anti-cholinergics, 66–67
Long-acting beta agonists, 66–67
Long-acting beta-2 agonists, 66–67
Lopressor and Toprol-XL (metoprolol), 104,
    105t
Lortab, 92
"Loss of results", 5–6

M

Madol, 15t
Ma-Huang plant (*Ephedra sinica*), 25
Marathon mice, 133
Masenate, 15t
Masking agents, 116t, 119–120
Mass spectrometry (MS), 115
Maximal expiratory flow–volume curve,
    73–74
Metanabol, 15t
Metandren, 15t
Metaproterenol, 66t
Methandrostenolone, 15t
Methitest, 15t
17α-Methyl testosterone, 13–14
Methyltestosterone, 15t
Milwaukee Journal Sentinel, 94
Mineralocorticoid receptors (MRs), 44–45
Mineralocorticoids, 38t, 44–45
MK-0773, 78t
Morphine, 91–92
    addiction, 96
Mu (μ) opioid receptor, 95t
Muscle mass, 33
    increasing, 14–16

Muscular contractility, 106
Myocardial infarction, 103
Myocardial ischemia, 72

N

Nandrolone, 15t
Naposim, 15t
Narcotics, 91, 116t
  addiction, 97–98
  arms of Morpheus, 91–92
  opioid system and pain, 94–96
  to relieve pain, 92–94
  substances, 96–97
National anti-doping agencies, 121, 130
National Collegiate Athletic Association (NCAA), 30
National sports agencies, 7
Neo-Hombreol, 15t
Nerve block clinics, 96
Neurotransmitters, 28, 68
Nociceptors, 93f
Nonsteroidal biochemical pathway, for induction of protein synthesis, 70f
Nonsteroidal cytoplasmic PI3K/Akt/mTOR/p70SK6 anabolic pathway activation, 68–71
Norbolethone, 15t
Norepinephrine, 28, 65, 68, 91, 95–96, 103, 106
Normodyne and Trandate (labetalol), 104, 105t

O

Operación Galgo (Operation Greyhound), 132
Opioid receptor subtypes, 94, 95t
Opioids, 91–92, 94, 97–98
  side effects, 98t
Opium poppy (*Papaver somniferum*), 91
Orciprenaline, 66
Oreton, 15t
Ormeloxifene, 78t
Ospemifene, 78t
Osteopenia, 72
Osteoporosis, 14, 72
Oxandrin, 15t
Oxandrolone, 15t
Oxycodone, 91–93, 96
OxyContin, 92, 97
Oxymetholone, 15t

P

Pain management, in athletes, 98b
Pain signals, 41–43, 94–95
Peptide hormones and related agents, 116t
Peptide–protein hormones, 49, 50t
  acromegaly, blood hyperviscosity, and death, 59
    erythropoietin, 59
    growth hormone, 59
  actions of, 56–58
    athlete biological passport and EPO, 58
    erythropoietin, 57–58
    growth hormone, 56–57
  increasing muscle mass and enhancing oxygen delivery, 52–53
    erythropoietin, 52–53
    growth hormone, 52
  offsetting endocrine defects and dysfunctions, 49–52
    erythropoietin, 51–52
    growth hormone, 49–51
  ramping up hormone actions within cells, 53–56
    erythropoietin, 54–56
    growth hormone, 54
Perandren, 15t
Percocet, 92
Percodan, 92
Performance enhancement, future of, 129
  binary gender constructs, 134–135
  doping, legalization and regulation of, 133–134
  doping deterrence, 130
  education of athletes, 130–131
  gene doping, 132–133
  sports doping, 135–137
  staying ahead of the doping athlete, 131–132
Performance-enhancing drugs (PEDs), 1–2, 8–9, 13, 21–22, 28, 37, 45–47, 49, 56, 65, 79, 95–96, 103
Peripheral analgesic action, 94–95
Peroxisome proliferatoractivated receptors delta (PPAR-δ), 132–133
Personal development, 8
Pethidine (meperidine), 91–92
Phenylpropanolamine, 26–28
Pheraplex, 15t
Physical dependence, 97
PI3K–Akt pathway, 55f
Plasma expanders, 119
Positive doping cases in Olympic Games 1968–2014, 5t

Positive test result, 121–124
Postsynaptic receptors, stimulation of, 28
Prader–Willi syndrome, 51
Prednisolone, 38*t*
Prednisone, 38*t*
Pressurized metered-dose inhaler, 65
Primary medical conditions, 8–9
Primary steroid hormones, 82*t*
Primum non nocere, 46*b*
Proellex, 78*t*
Progesterone, 82*t*
Proteogenesis, 14, 16–17
Pseudo-Cushing's syndrome and
        mineralocorticoid actions, 44–45
Pseudoephedrine hydrochloride (Sudafed),
        26–28
Psychedelics, 91
Psychosis, 91
Psychotomimetics, 91
Public disclosure of the violation, 124

R
RAD140 (Testolone), 78*t*
Raloxifene, 78*t*, 84
Receptor signaling, 16–17
Recombinant human EPO (rhEPO), 52, 61
Relaxation, 103
Rescue inhalers, 66–67
Resistance training, 71
Responders and nonresponders, 61*b*
Rewards, 7

S
S-22, 78*t*
Salbutamol, 66
Salmeterol, 66, 66*t*
Sample collection, 114*t*
Sarcoplasmic reticulum, 106
SDS-PAGE (sodium dodecyl sulfate-
        polyacrylamide gel electrophoresis),
        117–118
Secondary medical conditions, 8–9
Secondary messenger cascade mechanisms,
        28–30
Sectral (acebutolol), 105*t*
Selective androgen receptor modulators
        (SARMs), 77–80, 78*t*, 83–85
Selective estrogen receptor (ER) modulators
        (SERMs), 77–79, 78*t*, 84–85
    for breast cancer survivors, 86*b*
Selective progesterone receptor modulators
        (SPRMs), 78–79, 78*t*
Selective receptor modulator drug, 77–78

Serotonin, 43–44, 95–96
Short-acting beta agonists, 66–67
Skeletal muscle contractile proteins, 71
Skeletal muscle membrane integrity, 84
Soldier's disease, 96
Sports doping, 26–28
    in ancient civilization, 10*b*
    future of, 135–137
Sports ethics, 6–8
Sports injury, 46
Stanozolol, 15*t*
State-sponsored Russian doping scandal, 125*b*
Steroid hormone receptors, 16–17
Steroids. *See* Anabolic steroids
Stimulants, 25, 116*t*
    intoxication, 32
    losing weight, staying awake, gaining focus,
        25–26
    neurochemistry, 28–30
    performance enhancement evidence, 30–32
        amphetamine, 30–31
        caffeine, 31–32
    scheduled by the United States Drug
        Enforcement Agency, 27*t*
    speeding and crashing, 32
    staying athletically lean and on task, 26–28
    withdrawal, 32
Stromba, 15*t*
Strychnine, 1
"Substantial assistance", 122–124
Supervillin, 17
Supraphysiologic, 45
Sympathomimetics, 66
Synandrol, 15*t*
Syncope, 72
Systemic corticosteroids, 66–67

T
Tamoxifen, 78*t*, 79, 84, 86–87
Team sports, 7–8
Telapristone (CDB-4124), 78*t*
Tenormin (atenolol), 104, 105*t*
Terbutaline, 66, 66*t*
Testicular transplantation, 13
Testing of athletes, 1–2
Testolactone, 84
Testosterone, 13–14, 17, 78–80, 82*t*, 84–85
    anabolic effects of, 17–18
    doping, 120
    negative feedback modulation of, 83*f*
Testosterone cypionate, 15*t*
Testosterone enanthate, 13–14, 15*t*, 83–84
Testosterone propionate, 15*t*

Testosterone undecanoate, 13–14
Testostroval, 15*t*
Testoviron, 15*t*
Testred, 15*t*
Testrin, 15*t*
Testro LA, 15*t*
Tetrahydrogestrinone (THG), 15*t*, 118–119, 132
"The Clear", 15*t*, 118–119, 132
Therapeutic use exemption (TUE), 8–9
Tissue-selective drugs, 77–78
Toremifene, 78*t*
Tramadol, 95–96
Trenbolone, 15*t*
Triglyceride lipase (HTL), 20
Troponin, 106
Turner syndrome, 51

U
Ulipristal acetate ("Ella"), 78*t*
Ultandren, 15*t*
United States Anti-Doping Agency, 5–6
United States Drug Enforcement Agency (DEA), 26
United States National Institute on Drug Abuse (NIDA) statistics, 97–98
US Anti-Doping Agency, 81, 121
US Drug Enforcement Agency (DEA), 91–92

V
Vascular smooth muscle, 106–107
Vetanabol, 15*t*
Vicodin, 92
Victims of the East German Medal Machine 1970s to 1990s, 21*b*
Vilanterol, 66, 66*t*
Virilon, 15*t*
Visken (pindolol), 104, 105*t*

W
Weight loss and body composition, 33*b*
Whizzinator kit, 120
Winstrol, 15*t*
Women athletes, 137*b*
World Anti-Doping Agency (WADA), 2, 25, 31–32, 39, 53, 68, 72–75, 80, 91, 103, 130
and legal problems, 3–6
testing procedures, 113–115
WADA Prohibited List, 4–5
World Anti-Doping Code, 3–4, 113, 121

Y
YK11, 78*t*

Z
Zebeta and ziac (bisoprolol), 105*t*

Printed in the United States
By Bookmasters